Trade, Poverty and the Environment

Trade, Poverty and the Environment

The EU, Cotonou and the African-Caribbean-Pacific Bloc

Adrian Flint

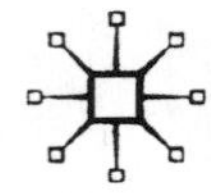

First published in 2008 by
PALGRAVE MACMILLAN
Houndmills, Basingstoke, Hampshire RG21 6XS and
175 Fifth Avenue, New York, N.Y. 10010
Companies and representatives throughout the world.

PALGRAVE MACMILLAN is the global academic imprint of the Palgrave Macmillan division of St. Martin's Press, LLC and of Palgrave Macmillan Ltd. Macmillan® is a registered trademark in the United States, United Kingdom and other countries. Palgrave is a registered trademark in the European Union and other countries.

ISBN-13: 978–0–230–51678–6 hardback
ISBN-10: 0–230–51678–5 hardback

This book is printed on paper suitable for recycling and made from fully managed and sustained forest sources. Logging, pulping and manufacturing processes are expected to conform to the environmental regulations of the country of origin.

A catalogue record for this book is available from the British Library.

A catalog record for this book is available from the Library of Congress.

10 9 8 7 6 5 4 3 2 1
17 16 15 14 13 12 11 10 09 08

Printed and bound in Great Britain by
CPI Antony Rowe, Chippenham and Eastbourne

For my family

Contents

List of Tables

Currency Conversions

For the sake of comparison, all figures expressed in currencies other than the US Dollar are accompanied by a Dollar conversion in parentheses. At the time of writing, the following exchange rates were in effect (April 2007):

Currency	Dollar ($) Value
Australian Dollar (AUS$)	0.77
British Pound (£)	1.9
Euro (€)*	1.3
South African Rand (R)	0.14

*The Euro was preceded by the European Currency Unit (ECU) which operated throughout the Polity between 1979 and 1999. When the Euro was introduced in 1999 it replaced the ECU at a 1:1 ratio. The ECU was itself preceded by the Monetary Unit of Account.

List of Abbreviations and Acronyms

ACP	African-Caribbean-Pacific
AMCOW	African Ministerial Conference on Water
ANC	African National Congress [South Africa]
ASEAN	Association of South East Asian Nations
CAP	Common Agricultural Policy
CAFOD	Catholic Fund for Overseas Development
CARICOM	Caribbean Community and Common Market
CARIFORUM	Caribbean Forum of African, Caribbean and Pacific States
CDM	Clean Development Mechanism
CEMAC	Central African Communauté Economique et Monétaire de l'Afrique Centrale
CFA	*Communauté française d'Afrique* [franc zone]
COMESA	Common Market for Eastern and Southern Africa
CU	Customs Union
DAWASA	Dar es Salaam Water and Sewage Authority
DCP	Data Collection Platform
DEAT	Department of Environmental Affairs and Tourism [South Africa]
DG	Directorate General [European Commission]
DME	Department of Minerals and Energy [South Africa]
DRC	Democratic Republic of Congo
DWAF	Department of Water Affairs and Forestry [South Africa]
EAC	East African Community
EBA	Everything But Arms
EC	European Commission
ECOWAS	Economic Community of West African States
ECSC	European Coal and Steel Community
ECFSP	European Commission Food Security Programme
EDF	European Development Fund
EEC	European Economic Community
EFSA	European Food Safety Authority
EIB	European Investment Bank
EMS	European Monetary System

EPA	Economic Partnership Agreement
EPRD	European Programme for Reconstruction and Development
ESA	Eastern and Southern African [Configuration]
EU	European Union
EUEI	European Union Energy Initiative
EUWI	European Union Water Initiative
EWS	Early Warning Systems
FAO	Food and Agriculture Organisation
FDA	[American] Food and Drug Administration
FDI	Foreign Direct Investment
FEWSNET	Famine Early Warning Systems Network
FTA	Free Trade Area
GATT	General Agreement on Tariffs and Trade
GATS	General Agreement on Trade in Services
GDP	Gross Domestic Product
GIEWS	Global Information and Early Warning System
GMO	Genetically Modified Organism
GNP	Gross National Product
GSP	Generalised System of Preferences
HDI	Human Development Index
HIPC	Heavily Indebted Poor Countries
HYCOS	Hydrological Cycle Observing System
IEA	Institute of Economic Affairs
IFI	International Financial Institutions
IMF	International Monetary Fund
ITDG	Intermediate Technology Development Group
LDC	Least Developed Country
LHWP	Lesotho Highlands Water Project
MDG	Millennium Development Goals
MDRI	Multilateral Debt Relief Initiative
MERCOSUR	*Mercado Comun del Sur* (Southern Common Market)
MNCs	Multinational Corporations
MTR	Mid-Term Review
NamWater	Namibian Water Corporation
NBI	Nile Basin Initiative
NGO	Non-Governmental Organization
NIC	Newly Industrialised Countries
NIEO	New International Economic Order
ODI	Overseas Development Institute

OECD	Organisation of Economic Cooperation and Development
OPEC	Organisation of Petroleum Exporting Countries
PBMR	Pebble Bed Modular Reactor
PIF	Pacific Island Forum
PPP	Public Private Partnership
PSE	Producer Support Estimates
PUP	Public Public Partnership
PV	Photovoltaic
RECs	Regional Economic Communities
RDP	Reconstruction and Development Programme [South Africa]
RRSU	Regional Remote Sensing Unit
SACU	Southern African Customs Union
SADC	Southern African Development Community
SADCC	Southern African Development Coordination Conference
SAMTRAN	Support to Agricultural Marketing Trade Negotiating [project]
SAMWU	South African Municipal Worker's Union
SAP	Structural Adjustment Programme
SAPP	Southern African Power Pool
SEA	Single European Act
SMO	Sugar Market Order
SPS	Sanitary and Phytosanitary [measures]
SSA	Sub-Saharan Africa
STABEX	*Système de Stabilisation des Recettes d'Exportation*
SYSMIN	*Système de Stabilisation de Recettes d'Exportation de* Produits Miniers
TAC	Technical Advisory Committee [AMCOW]
TCDA [EU-SA]	Trade, Cooperation and Development Agreement
TDCA	Trade, Development and Cooperation Agreement
TEU	Treaty of European Union
UNDP	United Nations Development Programme
UNESCO	United Nations Educational, Scientific and Cultural Organisation
UNFCCC	United Nations Framework Convention on Climate Change
USAID	United States Agency for International Development

WAEMU	West African Economic and Monetary Union [with UEMOA as its French acronym]
WFP	World Food Programme
WHYCOS	World Hydrological Cycle Observing System
WMO	World Meteorological Organization
WSS	Water and Sanitation Services
WSSA	Water and Sanitation Services South Africa
WSSD	World Summit on Sustainable Development
WTO	World Trade Organisation
WUC	Water Utility Corporation [Botswana]
WWF	World Wildlife Fund

Acknowledgements

This book is dedicated to my family in South Africa, thank you for everything.

This work was a long time in the making and I am grateful to all those who contributed along the way: the staff of the Politics Department at Rhodes University (South Africa); my officemates at the University of Dundee, John MacDonald and Abdullah Yusuf; and all the students I have taught over the years. I would specifically like to thank Norman MacQueen and Brian Baxter for their guidance and encouragement whilst I based in Scotland. Furthermore, I am grateful to all of those at the University of Dundee who generously offered me advice and the benefit of their experience, Richard Dunphy in particular. At the University of Bristol, I would like to thank Sarah Childs, Vernon Hewitt, Tim Edmunds and Eric Herring for their recommendations and support.

Last, but by no means least, special thanks must go to Jill Payne, who has selflessly read just about everything I have ever written, thank you very much, this book could never have been completed without your input.

Introduction

According to the central provisions of the Cotonou Agreement, the European Union (EU) is committed to facilitating 'poverty eradication, sustainable development and the gradual integration of the African-Caribbean-Pacific (ACP)[1] countries into the world economy' (Cotonou Partnership Agreement, 2000, Preamble). This volume evaluates the EU's commitment to the spirit of Cotonou. Following the demise of the 25-year-old Lomé regime, the new partnership agreement was signed between the two groups in Cotonou, Benin, on 23 June 2000, heralding a new era in a long-standing relationship. Within the Treaty, EU policy-makers make frequent linkages between the terms 'poverty alleviation' (or eradication) and 'sustainable development' (Articles 1, 9, 19, 23, 24, 34, 73). It is this two-pronged approach to development that will form the basis of this study.

Poverty is a multifaceted concept and definitions abound.[2] Likewise, there are a number of poverty-combating strategies on offer. The prevailing development orthodoxy, centred largely on neoliberal macro-economic practices, is promulgated by all of the major international donors and financial institutions and the EU is no exception in this regard. The evidence in favour of this approach is, at best, mixed. However, it is not simply the neoliberal development paradigm but the manner of its implementation that demands scrutiny. Here, it is argued that the EU has adopted a stubbornly 'ideological' approach to development which is, at times, either inflexible or inappropriate or both.

The EU-ACP relationship does not take place in a vacuum. Instead, the relationship must be viewed, initially, as part of the broader framework of international development initiatives, sustainable development and poverty alleviation strategies. Arguments pertaining to both poverty alleviation and environmental degradation, as well as globalisation, are frequently emotive, heavily polarised and oversimplified. Anti-globalisation protestors, environmental groups and international financial institutions (IFIs) rarely see eye to eye. The reality, as is so often the case, is far more complex than many would admit. Herein lies the quandary. There are no easy solutions to these problems and likewise, there are no clear-cut heroes or villains. Instead, it is important for both sides to acknowledge the multifaceted nature of the issues confronting those living in extreme poverty, and to act accordingly.

I do not attempt to apportion blame for poverty and environmental degradation in developing countries solely on either the developed world in general or the EU in particular. It cannot be denied that domestic patterns of landholding, corruption and wasteful practices have all served to place undue pressure on people and their natural environments in developing countries. However, whilst cognisant of the fact that such countries should have a responsibility to keep their own 'houses in order', this line of argument is not pursued here. Instead, the study concentrates on the impact of EU policies and the implications of these where the promotion of sustainable development and poverty alleviation for the whole of the ACP bloc is concerned.

The sustainable development concerns facing low-income countries differ from those of richer countries. In the EU, most environmental degradation occurs as a result of high levels of personal consumption. In developing countries it is the opposite. The global impact of activities in developed countries is more severe in an aggregate sense than in developing countries when, for example, greenhouse gas emissions are considered, while environmental degradation in developing countries takes a potentially more immediate toll on local populations. The environmental consequences associated with high levels of poverty – soil erosion, desertification, deforestation and biodiversity loss – are ones which are often overlooked but which must be addressed if people's lives are to be improved. This is especially true for those in rural areas. The deprived in rural areas are very much at the mercy of their local surroundings for their day-to-day subsistence and well-being, and any meaningful attempt at development must incorporate environmental considerations. As nearly 70 per cent of people living in extreme poverty live in rural areas, any real attempt at poverty reduction must start in this sector (European Commission 2000a).

The links between environmental factors and basic subsistence are immediately apparent – disease in Africa is largely environment-related, as are high rates of infant mortality and low life expectancy (Steele *et al.*, 2002, p. 15). With little access to clean water, adequate sanitation, reliable energy supplies or food security, those on very low incomes tend to be susceptible to a number of environmental stresses. However, as they are overwhelmingly dependent on their natural surroundings for their day-to-day subsistence, they are often forced to degrade or exploit the natural environment in order to support themselves. Since the majority of the deprived in rural areas are incapable of deriving subsistence from purely agricultural sources, they are invariably forced into finding alternative means of generating income,[3] many of which, such as charcoal production, are environmentally destructive. The effects of such actions are often compounded by the fact that the majority of those living in extreme poverty also live in ecologically sensitive areas, having been pushed onto marginal land (Chisholm 2001, p. 12). This is particularly true for Africa, which is the continent with the world's highest proportion of people living in extreme poverty (World Bank 2004). In this respect, environmental degradation on an acute scale puts the livelihoods of those living in poverty at risk due to inevitably diminishing returns, thereby ensuring that the 'poverty trap' cannot be broken. Sustainable development strategies in countries such as those of the ACP must therefore focus on environmental protection as part of any poverty reduction programme, particularly in rural areas, if poverty is to be curtailed.

Being able to provide for basic subsistence is clearly the platform for any poverty alleviation strategy, and analysis of the 'poverty trap' provides a rationalisation for including environmental protection strategies within that platform. The devil, however, is in the details, given that there are a plethora of policy frameworks that claim to represent paths forward. The EU bases its stance on the assumption that its neoliberal poverty alleviation strategies will result in both a cessation of poverty-related environmental degradation and an increase in economic growth. This study challenges this assumption from three basic perspectives. First, it considers the evidence behind the equation of economic growth with lower levels of environmental degradation. Second, it evaluates the impact of neoliberal economic strategies on economic growth and poverty reduction. Finally, it discusses the extent to which, even if successful in theory, it is possible to weld environmental considerations to this particular model of development. Poverty alleviation, far from evolving in tandem with environmental protection, is on the one hand endangered by it, in the form of environmental tariffs, and yet on the other still threatened by the lack of it, in that neoliberal economic strategies continue to emphasise environmentally unsound cash-cropping.

Development has long centred on a seemingly simple debate, the crux of which revolves around two basic positions: whether to promote self-sufficiency or self-reliance.[4] Self-sufficiency entails people having access to suitable agricultural land and necessary natural resources in order to subsist. Self-reliance on the other hand necessitates adequate levels of employment and purchasing power so as to be able to acquire an adequate standard of living. Both positions involve a number of assumptions and risks. Self-sufficiency is often associated with heavy state involvement in an era currently governed by an increasing 'roll back' of the state. Self-reliance strategies, on the other hand, are usually tied to neoliberal policies such as trade liberalisation and the participation of developing countries in the global economy. Environmental considerations are important when assessing both sides of the debate as such strategies must be sustainable, both economically and environmentally, if there are to be long-term positive benefits. It is self-reliance strategies that are generally espoused by IFIs such as the International Monetary Fund (IMF) and World Bank, as well as international donors such as the EU. Given that this is the case, this study will concentrate largely on the efficiency of EU policies aimed at promoting this form of development.

It is argued that there remains a place for state invention and subsidies, as well as a need for special measures to protect developing countries from 'dumping' and cheap imports. The record of state intervention in stimulating economic growth in developing countries is not, on the whole, a proud one, which is why such policies should not be adopted on an *ad hoc*, unilateral basis. Rather, they should be locked into the economic reforms being pressed for by international donors such as the EU. Such precautions would ensure 'good governance' principles and a defence against 'rent seeking' activities which have been all too ubiquitous in the past. However, the EU appears determined to press for self-reliance strategies regardless of their suitability. The result is a potential increase in poverty levels.

Theories of sustainable development

It is important at this point, for the sake of clarity, to introduce an overview of the debate surrounding the concept of sustainable development. Despite the contested nature of the term, sustainable development has become a popular 'catch phrase' since first being introduced to the wider public following the publication of the 1987-Bruntland Report. The commonly cited definition of the term, as set out in the Report, states that sustainable development entails 'development that meets the needs of the present without compromising the ability of future generations to meet their own needs' (World Commission on Environment and Development 1987, p. 43).

Yet, despite the popularity of the Bruntland definition there remains little precise consensus as to what the above actually entails in terms of economic policy, norms and values. 'Sustainable development' is a phrase that is regularly invoked by a variety of interests that ordinarily would have little in common, from NGOs such as Oxfam and Greenpeace, to the British high street chain Marks & Spencer. The vague and ill-defined nature of the term provides it with a malleability that has engendered it with widespread appeal and left it open to a range of interpretations, especially at the theoretical level. Herman Daly (1992, 1996), former World Bank analyst and ecological economist, conceptualises sustainable development in a manner that differs markedly from that of more orthodox environmental economists such as David Pearce (1993) or such 'green dissenters' as Julian Simon (1981, 1996), Matthew Ridley (1995, 1996) and Bjørn Lomborg (2001).

Despite the breadth of views, proponents of sustainable development can be roughly divided into two separate camps, one favouring a 'weak' interpretation of the term and the other favouring a more radical agenda. The advocates of a 'strong' interpretation of the term are epitomised by the likes of Daly (1996, p. 3), who argues that 'production growth must not push us beyond the sustainable environmental capacities of resource production and waste absorption... once that point is reached, production and reproduction should be for replacement only. Physical growth should cease, while qualitative improvement continues'. On the other hand, those advocating a 'weak' interpretation of the term can be broken down into two further subsets. The first, exemplified to a certain extent by Simon, Ridley and Lomborg, argues that the 'cure' for environmental degradation is often worse than the 'disease' and furthermore that, as Lomborg (2001, p. 5) states, 'our problems are getting *smaller* and not bigger, and... frequently the offered solutions are grossly inefficient'. What is important is that the level of 'total' capital does not diminish and that economic growth is not curtailed. The growing popularity of such views is evidenced by their deployment in fictional 'eco-thrillers' such as Michael Crichton's *State of Fear* (2004). A somewhat less complacent interpretation of weak sustainability is provided by environmental economists such as Pearce, who proffer solutions on the basis of market-driven programmes and a need to 'get prices right'. Pearce (1993, p. 3) argues that until 'the economic value of environmental quality is an everyday feature of the way we compute progress and, more importantly, the way we make economic decisions then ... the environment will not be given a fair chance'.

It is important then, given this disparity of views, and the importance of such an understanding in guiding policy formulation, to ascertain how the notion of sustainable development is interpreted by the EU. This study demonstrates the similarity of the EU's position to that of environmental economists such as David Pearce, and thus its compatibility with the neoliberal economic paradigm embodied by such institutions as the World Trade Organisation (WTO), the IMF and the World Bank. The Cotonou Agreement's bluntly orthodox credentials are on display in Articles 34, 36–37, 39, 46–48 of the Treaty in which it is stated that 'economic and trade cooperation shall be implemented in full conformity with the provisions of the WTO' (Article 34).

Theories of North–South relations and development

The formulation of ties between the EU and the ACP must be placed within a wider global and historical context. No relationship of this nature takes place in isolation and it can only be understood against the backdrop of the global political economy. Growth theory, modernisation theory, neoliberalism, dependency theory and world systems theory have all sought to explain the unequal relationship between North and South. All of these theories are largely prescriptive as well as explanatory, and each suggests a specific course of action. The triumph of the neoliberal economic paradigm over its competitors in recent decades has resulted in the supremacy of the idea of trade liberalisation being the path to greater development and self-reliance. As key players in the global economic system, the EU and its component member states have worked hard to cement neoliberal orthodoxies within their relations with less-developed countries.

The oil shocks of the 1970s, together with the collapse of the Bretton Woods economic system and the worldwide economic recession of the early 1980s, heralded the onset of a crippling debt crisis in the South from which, even now, many countries have not fully recovered. The end of the Cold War further eroded the influence of developing countries within the international arena. It is against this background that the current state of relations between the EU and the ACP must be addressed.

As stated, any assessment of EU policy reveals a strong neoliberal stance in relation to development. Private sector involvement, deregulation and liberalisation are all heavily promoted by the EU as core components of any development strategy. Many have questioned the viability of such strategies in delivering tangible benefits for those living in extreme poverty, given that the majority of such people are poorly placed to take advantage of the opportunities afforded by such an approach. Liberalisation frequently results in the 'dumping' of agricultural produce on developing nations' markets, thereby undermining already precariously balanced local rural sectors. Moreover, the privatisation of state utilities has been shown to lead to significant price increases for essential services such as drinking water. This in turn has resulted, in some areas, in the return of nineteenth-century epidemics such as cholera – the result of people being forced to resort to unsafe but cheaper alternatives (Bayliss 2002; Corporate Europe Observatory 2003; McDonald 2002; Pauw 2004; Santoro 2004). In addition, there is a significant degree of debate as to whether market-based solutions will result in any environmental gains. However, and of more fundamental concern, it is frequently argued that even if the EU is correct in its assumptions regarding development policy, EU domestic policies in effect negate any possibility of such strategies being effective.

A review of the evidence

The cornerstone of the Cotonou regime is the establishment of WTO-compatible Economic Partnership Agreements (EPAs) between the EU and its ACP partners. These agreements are essentially reciprocal Free Trade Agreements (FTAs) and entail the liberalisation of 'substantially all' trade between the EU and six regional groupings of ACP countries. The negotiations surrounding the creation of these EPAs were initiated in September 2002 and are set to culminate on 31 December 2007 (although, at the time of writing, this deadline appears increasingly unlikely). Talks between the two sides have brought the development debate sharply into focus, highlighting the enormous disparity in power between the North and the South. There is a wealth of literature outlining the pace and development of these negotiations. The European Commission (EC) has produced a number of policy papers in this area, together with sustainability assessments. The World Bank has also produced extensive material pertaining to poverty alleviation, regionalism and sustainable development. In the book, this material is frequently juxtaposed against material from development NGOs such as Oxfam, Catholic Fund for Overseas Development (CAFOD), Christian Aid and Tearfund. I also draw on material from various UN agencies such as the United Nations Development Programme (UNDP) and Food and Agriculture Organisation (FAO) as well as 'thinktanks' such as the Overseas Development Institute (ODI) and the European Centre for Development Policy Management (ECDPM). Analysis of the EU's commitment to the three stated objectives inherent in the Cotonou Agreement is thus based on an extensive review of existing literature in this area.

An anthropocentric approach

This study approaches environmental issues from a purely anthropocentric view, focusing on the impact of environmental degradation on the poor in developing countries. This is not to discount other approaches arguing for environmental protection such as those expressed by 'Deep Ecologists'[5] Arne Naess (1973), George Sessions and Bill Devall (1985). Rather, it is a deliberate attempt to narrow the focus of the debate and, more specifically, to avoid advancing an overly normative/utopian argument in a field heavily dominated by empirical analysis. While normative arguments have their place, pushing the parameters of debate and challenging established views, such arguments often tend to remain peripheral to the policymaking process. Thus, whilst arguing against complacency, the study can be viewed as an attempt to both highlight deficiencies in EU-ACP relations and to suggest how these might be addressed within the current system.

Overall structure

Chapter 1: The EU and the ACP – from Yaoundé to Cotonou

Chapter 1 contextualises the EU-ACP relationship; the Yaoundé Conventions, the Georgetown Agreement, the Lomé Conventions and the build-up to the Cotonou Agreement. The newfound importance of regionalism in the EU's relations with developing countries is stressed, with a focus on how regionalisation has come to be the platform for EU development policy in the post-Lomé era of cooperation.

Chapter 2: Towards an EU Definition of Sustainable Development

This chapter outlines the differing interpretations of sustainable development and the potential ramifications of this debate for policymakers. It provides a review of the arguments and strategies posited by different interests, and places the EU's own interpretation of sustainable development within this wider debate.

Chapter 3: The Rise of the Neoliberal Consensus

Chapter 3 positions the EU-ACP relationship against the backdrop of increasing globalisation and broader North–South relations. In this respect, the relationship is treated as part of a wider study involving the historical interaction between richer and poorer nations. The chapter contends that increasing levels of globalisation do not necessarily indicate a set course of action where the promotion of development in poorer regions is concerned.

Chapter 4: The 'Everything But Arms' Initiative

The Cotonou Agreement remains in its relatively early stages and a realistic assessment of its impact on sustainability is difficult to assess before the implementation of its EPAs. However, the Everything But Arms (EBA) initiative, aimed at Least Developed Countries (LDCs), is already in place. A range of conclusions relevant to EPAs can therefore be gleaned from a consideration of the EBA: its success in promoting environmentally sustainable, pro-poor growth amongst the ACP-LDCs can offer a number of insights into the viability of sustainable development targets in the remainder of the ACP bloc. The EBA initiative thus provides a test case for EU poverty reduction strategies in the developing world and a possible foretaste of the post-2008 post-Lomé era.

Chapter 4 is followed by a detailed examination of four key areas of EU development policy targeting sustainable development and poverty reduction amongst its ACP partners: Chapter 5 discusses trade policy, Chapter 6 discusses agriculture and food security and Chapters 7 and 8 discuss development strategies relating to energy and water provision respectively. These four sectors are useful points of departure because they highlight a number of issues pertaining to the impact of EU programmes and policies in the South.

Chapter 5: Trade Policy, Poverty Alleviation and Sustainable Development

The EU has steadfastly argued that increased levels of trade will be of significant benefit to the ACP countries and will help to facilitate environmentally sound, socially aware development. In conjunction with the IMF, World Bank and other neoliberal institutions, the EU has actively encouraged the ACP countries to liberalise their economies and to embrace global markets. Chapter 5 argues that, even if the EU is correct in this view, and that neoliberal economic principles are indeed the solution to poverty in the South, the EU negates any possible positive effects of such economic restructuring by refusing to fully liberalise its own markets.

Chapter 6: Agriculture, Food Security and Rural Development

This chapter centres on the EU's attempt to ensure agricultural prosperity and adequate food levels in the ACP countries. As the majority of the world's poor live in rural areas, it is imperative that, if poverty is to be eradicated, then more attention be devoted to the rural/agricultural sector. Here, the view is posited that EU-sponsored programmes in these regions, put in place to prevent famine and insecurity, have frequently failed. Moreover, the EU's continuing reluctance to reform its Common Agricultural Policy (CAP) means that some ACP farmers will continue to struggle to profit from their relationship with the EU.

Chapter 7: Energy Provision, Poverty Alleviation and Sustainable Development

Chapter 7 addresses non-agricultural aspects of development, specifically energy provision. The EU stated its commitment to targeting 'energy poverty' at the World Summit on Sustainable Development (WSSD) held in Johannesburg, South Africa, in 2002. The link between raised levels of energy consumption and improved living standards is not always fully appreciated. However, the associated environmental costs of such increased provision can be high. EU initiatives in this area are thus evaluated in terms of their commitment, in practice, to sustainable development principles.

Chapter 8: Access to Water and Effective Sanitation – Provision and Security of Supply

As with increased per capita levels of energy consumption, functioning water and sanitation services are crucial to alleviating poverty.

Approximately, three million people per year die from waterborne disease, the vast majority of whom are children. Poor sanitation exacerbates this problem. At the WSSD, the EU committed itself to bolstering water and sanitation services amongst the ACP countries. However, EU projects in this area have tended to involve the privatisation of such services – not always with positive results. In addition, this chapter considers issues pertaining to water security, including the extent to which the EU is endeavouring to provide such protection.

Chapter 9: EU-ACP Economic Partnership Agreements

This chapter outlines what Cotonou entails, the extent to which it differs from Lomé and the likely outcome of this new regime in terms of poverty alleviation and the integration of the ACPs into the global economy. The new treaty moves away from the non-reciprocal elements inherent within Lomé, with its tacit admission that, in trading terms, not all countries should necessarily be treated equally (a form of positive trade discrimination), towards a system whereby all countries, with limited exceptions, are treated equally, be they developed or developing. Chapter 9 focuses on the proposed EPAs put forward in response to complaints regarding Lomé's breaches of the General Agreement on Tariffs and Trade /World Trade Organisation (GATT/WTO) international trading regime.

Conclusion

On the basis of its own targets and criteria, the EU's ACP development policies promote neither poverty alleviation nor sustainable development. EU funding does not adequately target the poor, and, moreover, does little to offset the environmental degradation usually associated, whether directly or indirectly, with high levels of poverty. Much of the blame for this failure can be found in a procrustean tendency to stress the efficacy of market-led strategies, as witnessed by the EU's ongoing insistence on privatisation and Public–Private Partnerships (PPP) in the region, despite the lack of clear evidence that such options represent the best way forward.

1
The EU and the ACP – from Yaoundé to Cotonou

Introduction

The evolution of EU-ACP cooperation between 1957 and 2000 is significant in that it contextualises the European Commission's current attempts at promoting development amongst its ACP partners. Cooperation between the EU and the countries that came to form the African-Caribbean-Pacific (ACP) bloc in 1975 is almost as old as the Polity itself. French influence in the early days of the EU ensured the development of a 'special relationship' between the newly formed European Economic Community and countries tied to Europe through colonialism. The special nature of this relationship came to be entrenched with the signing of the Georgetown Agreement in 1975, the formation of the ACP bloc and the negotiation of the Lomé Convention. The Lomé Convention and its later incarnations were to govern EU-ACP cooperation for 25 years and formed the heart of the EU's development agenda. The unique bond between the EU and the ACP was epitomised by the Lomé trading regime which established non-reciprocal, privileged access to EU markets for ACP exports. However, by the 1990s it was clear that Lomé had run its course. The EU made it clear that the agreement would not be extended beyond 2000. A new partnership agreement, signed in Cotonou, Benin, in 2000, set out the blueprint for future EU-ACP relations. The Cotonou Agreement was a disappointment for many ACP countries. Despite the rhetoric, it appeared that the special relationship that had existed between the two blocs no longer exists. The EU's apparent 'abandonment' of the ACP can be explained by virtue of a number of factors: the end of the Cold War; the incompatibility of Lomé with the WTO trading regime; EU expansion; and the development of a neoliberal consensus by the major aid donors.

Contextualising EU-ACP cooperation

France had made it a condition of its accession to the Treaty of Rome that that there be some accommodation made for its colonial and ex-colonial territories. Consequently, a regime of association was created in 1957 and in 1958 the first European Development Fund (EDF) was established with a budget of 58 million European Units of Account ($58 million).[1] A significant proportion of this funding was distributed in the form of grants to French overseas territories. The relationship was formalised in 1963 by the signing of an association agreement in Yaoundé, Cameroon, between 18 newly independent francophone African states and the EEC. The Convention, which was subsequently renegotiated, was operative between 1963–1969 (Yaoundé I) and 1969–1975 (Yaoundé II), and established the basis for EU-ACP cooperation. The aim of Yaoundé was to enable these newly independent states to achieve significant economic development and thereby economic autonomy. In light of the proposals advanced by the Cotonou Agreement (see below), it is noteworthy that the Yaoundé regime stressed the need for reciprocity and non-discrimination in exchange for EDF grants. Furthermore, it actively encouraged the formation of regional partnerships so as to facilitate the creation of a Europe-Africa free trade zone (Lecomte 2001, p. 8). However, the Yaoundé goal of an EEC-Africa free trade zone went unrealised due to three key factors:

1. The majority of newly independent African states chose to base their economies on import-substitution strategies, which entailed high levels of protectionism. There was thus little enthusiasm for reciprocal trade links with the EEC.
2. French business interests, which had enjoyed preferential access to these markets, were anxious to ensure that other European companies did not undermine their security.
3. The United States was determined to prevent Europe claiming Africa as its sphere of influence and argued strongly against the concept.

When Britain was finally admitted to the EU in 1973, it was clear that a new, more effective agreement would need to be negotiated in order to accommodate Commonwealth interests. The EU decided to divide the Commonwealth countries into two groups, the 'associables' and the 'non-associables'. The associables included the African, Caribbean and Pacific Commonwealth members and these countries were invited to negotiate association agreements with the EU. The non-associables, such as India, were deemed too developed to warrant inclusion in such an agreement and were thus restricted to negotiating standard trade agreements with the Polity (Brown 2002, p. 45).

To the surprise of the EU, when talks began in July 1973, the ACP countries decided to negotiate as a bloc rather than in regional groupings (Hall and Blake 1979, pp. 111–25). This show of unity led to the formation of the ACP group in June 1975, when 46 developing countries signed the Georgetown Agreement. The fact that these countries chose to negotiate an association agreement as a single group significantly bolstered their leverage during the talks. Consequently, the resulting convention, Lomé I, appeared to offer ACP countries a number of favourable concessions. The demands for reciprocity and Free Trade Zones inherent in Yaoundé were dispensed with in favour of non-reciprocal tariff preferences for ACP countries. Furthermore, the EU introduced compensation mechanisms such as STABEX and SYSMIN[2] in order to help offset commodity price instability. Following the 1973 oil crisis, the ACP countries were able to extract such concessions from the EU as a result of European anxieties as to the continuing availability of raw materials and as a result of Cold War geo-political considerations. Furthermore, the negotiations took place in the wake of the UN call for a New International Economic Order (NIEO) in 1974, which demanded developed countries make fair the terms of trade between North and South. The first Lomé Convention was duly signed in 1975 between the then nine member states of the EU and the 46 ACP states. Formalising the group as it did, the Georgetown Agreement created the structures necessary to facilitate the agreement, including a secretariat. The Convention was subsequently renegotiated at five-yearly intervals, in 1980 (Lomé II), 1985 (Lomé III), 1990 (Lomé IV) and finally in 1995 (Lomé IV-bis).

Lomé was thus very different from Yaoundé. The agreement was a clear breach of the Most Favoured Nation (MFN) principles inherent in the General Agreement on Tariffs and Trade (GATT) and consequently required a waiver. Article I of the GATT demands that any trade concessions offered by one signatory to another must, in turn, be offered to all other signatories. The GATT does broadly allow for two exceptions to this rule, one, if the concessions are motivated by development concerns and two, if the concessions relate to the functioning of free trade areas. However, the GATT rules stipulate that concessions cannot discriminate. For example, if the concessions were offered to developing countries, then they would have to be available to all developing countries. Lomé trade preferences, on offer only to ACP countries, were thus a breach of the GATT. Moreover, running counter to many of the free trade aspects of the Yaoundé regime, Lomé offered banana, beef and veal, rum and sugar protocols which, in essence, guaranteed producers prices well above global market levels. Moreover, 'stabilisation' measures such as STABEX and SYSMIN further sidelined market pressures in the EU-ACP relationship.

Of all the ensuing accords, Lomé I was the closest to an agreement between partners. There was little or no conditionality and ACP members were free to formulate their own economic policies without undue outside interference. However, each successive renegotiation of Lomé diminished the power of the ACP within the relationship. The debt crisis of the 1980s, the collapse of world commodity prices and the end of the Cold War eroded its bargaining power, with the result that the EU was able to incorporate conditionality into later versions of the Treaty. Poverty reduction, good governance, gender equality and sustainable development all became hallmarks of Lomé's later incarnations (Dearden 2002, pp. 4–6). Positive as these conditions arguably are, it is nonetheless possible to view the relationship between the EU and the ACP as having become increasingly one-sided. The concessions secured by the bloc in the negotiations in the build-up to Lomé I were progressively 'rolled-back' by each successive renegotiation of the convention. Yet despite such increasing disparities, the majority of ACP countries were dismayed when it became apparent that the convention would not be renewed after the expiry of Lomé IV-bis in 2000.

The eroding of a special relationship

Lomé I represented the 'high water mark', from an ACP perspective, in EU-ACP relations. A true sense of partnership prevailed; the lack of conditionality allowed ACP countries to determine their own domestic economic policy together with systems of governance, cementing the 'uniqueness' of the relationship. However, the equilibrium of the partnership was rapidly undermined by changes in the global arena. The debt crisis of the 1980s, brought about by the oil crises of the 1970s and a collapse in global commodity prices, dramatically altered the balance of power between the two blocs. The international community's response to the debt crisis is generally described as the 'Washington Consensus' (see Chapter 3). The debt crisis saw a shift away, on the part of major donors, from project finance to the funding of 'policy reform' and the implementation of neoliberal economic practices. This shift in focus was spearheaded by the IMF and World Bank but rapidly adopted by other key actors such as the EU (Brown 2004, pp. 19–20). Chris Patten, Overseas Development Minister in 1988, succinctly articulated the view in Europe at the time: 'it makes no sense to argue one course in Brussels and other in Washington' (cited in Brown 2004, p. 22). The imposition of economic conditionality put the EU-ACP relationship under a degree of strain because it was incompatible with the partnership ethos that was supposedly at the heart of Lomé. However, the Commission maintained that the structural adjustment called for by the Washington Consensus was unavoidable and duly incorporated economic conditionality into the Lomé framework.

When it became clear, towards the late 1980s, that the structural adjustment policies advocated by the IMF and World Bank were having little effect in bolstering economic growth in developing countries, particularly those in Africa, the strategy was reconsidered. It was argued by advocates of adjustment that any failure in promoting development was as a result of poor implementation rather than any inherent policy deficiency. Consequently, the IMF and World Bank began to demand political as well as economic conditionality. It was believed that transparency and good governance would reduce corruption and would generate the necessary political will to force through reforms (Brown 2004, pp. 25–31). The EU rapidly followed suite, making good governance an important plank in its development strategy.

Once again, by the mid-1990s it was apparent that the majority of developing countries were underperforming economically, despite the introduction of economic and political conditionality. The international consensus shifted towards 'pro-poor' strategies aimed at promoting poverty alleviation. However, it was made clear from the outset that any poverty alleviation strategies devised would not be permitted to conflict with the neoliberal reforms also demanded by key donors such as the IMF and World Bank. The EU, once again, rapidly fell in line with the IMF/World Bank view of development. The language of poverty alleviation is clear to see in the 2000 Cotonou Agreement, just as economic governance was explicitly entrenched in Lomé IV and good governance in Lomé IV-bis. By 2000, the approach adopted by the EU to its ACP partners was a long way removed from that negotiated in 1975. Furthermore, instead of being a partnership based purely on the needs of the participants, the relationship became increasingly 'normalised' within a growing international development consensus.

It was not only the evolution of the Washington Consensus that was to affect the relationship between the EU and the ACP. The fall of the Berlin Wall in 1989 and the end of the Cold War were also to have a profound effect on the 'special nature' of the relationship. The end of the Cold War meant that, from a strategic perspective, African countries ceased to be as important, geopolitically speaking, as they once had been. Furthermore, it meant that markets and trading partners in Eastern Europe, long cut-off from the EU by the 'iron curtain', now became accessible. The EU's attention became increasingly focused on its near neighbours, to the detriment of its ACP partners. In 1970–74, out of the top 15 recipients of EU aid, 13 were ACP members. By 1996–97 this figure had fallen to just two, the remainder being made up of East European countries and countries bordering on the Mediterranean (Smith 2004, p. 62). The EU's commitment to its near neighbours was confirmed in 2004 when ten, mainly former communist bloc countries, became members of the Polity, swelling its membership to 25 and again, in 2007, with the accession of Bulgaria and Romania. Moreover, the end of the Cold War also served to open up other regions to European interests. Latin American countries, who were lying strictly within the American sphere of interest during the Cold War, have entered into a variety of arrangements with the Polity including an association agreement between the EU and MERCOSUR (Argentina, Brazil, Paraguay and Uruguay) in 1996 and a EU-Mexico free trade agreement in 1999. Negotiations for a MERCOSUR-EU free trade agreement are underway. The EU is similarly attempting to expand its influence in Asia, expanding on a cooperation agreement signed with the Association of South East Asian Nations (ASEAN) in 1980. In short, the EU has become a global actor since 1989 and the ACP countries are no longer the Polity's only development partners. Given this change in circumstances, it is possible to argue that Cotonou represents the 'rationalisation' of EU development policy with regards to the ACP. The ACP bloc is now simply one amongst many, no longer 'special' (Smith 2004, p. 62). By the time Lomé expired, the Generalised System of Preferences (GSP) regime that governed EU relations with non-ACP developing countries had been extended and enhanced to such a degree that the tariff differences between the two arrangements had fallen to just 2 per cent of each other (Dickson 2004, p. 42).

However, the simplest explanation for Lomé's demise was its incompatibility with the international trading regime enshrined in the GATT and later the WTO. The Lomé regime found itself under sustained pressure from 1994 onwards when a GATT Panel ruled that the non-reciprocity elements contained within the convention, as well as its discriminatory nature, meant that it was incompatible with the multilateral trading system. Consequently, Lomé IV-bis (1995–2000) required a WTO waiver before it could be implemented. Largely as a result of the GATT Panel's findings, the EU decided that any future EU-ACP agreement would have to be consistent with WTO rules.

It can be argued that it was becoming increasingly apparent to EU policymakers that certain 'sacred cows', such as the CAP, were also coming under sustained pressure to be reformed. The ACP regime had been relatively cheap to run and involved little in the way of actual costs to the EU. However, with the CAP under threat from outside interests such as the USA and WTO, it seems likely that the EU viewed an additional battle over preferences to third countries as being too 'costly' (Dickson 2004, p. 56). This change in attitude was apparent from the mid-1990s, epitomised by the EU's dealings with the newly democratised South Africa. The FTA negotiations between South Africa and the EU in the latter part of the 1990s provided observers with an indication of the Polity's vision of a post-Lomé agreement. From the outset it was made plain to the South African Government that any agreement would have to conform to WTO guidelines, with the result that the South African delegation was confined to negotiating within very narrow parameters (Gibb 2002, p. 120). It took two years of often acrimonious talks before an agreement was reached.

However, the EU is somewhat disingenuous when it implies that the WTO lies outside of its sphere of influence. The WTO is important, to a large extent, because the EU and EU members wish it to be so. The WTO is part of a regulated international system devised and supported by the major economic powers. EU-ACP relations are thus determined by the WTO because the EU has a large stake in the success of the regime. The influence of EU member states in other international financial institutions is also marked; 11 out of 24 executive directors in the World Bank are drawn from EU members. Likewise, EU members control 30 per cent of the vote in the IMF and contribute 7 out of the 24 directors (Bailey 2002, p. 5). Clearly, then, it is a case of the EU having chosen to abandon Lomé rather than having been forced to do so by powers outside its control.

Yet, despite the shift away from non-reciprocal exchange, the EU has been most anxious to 'talk up' the Cotonou Agreement in an effort to portray itself as the guardian of the poor and the marginalised. Poul Nielson, the then European Commissioner for Development and Humanitarian Aid, stated that the partnership would

> give a new momentum to the relationship between the ACP States and the European Union. It represents an important component of

> international efforts aimed at promoting sustainable development and reducing poverty... It has been achieved thanks to a shared political willingness to give an ambitious response to these problems, building on the *acquis* of twenty-five years of cooperation under the Lomé Conventions. (*The EU-ACP Courier* 2000, p. 2)

Notwithstanding the unequal balance of power between the EU and the ACP countries by the late 1990s, the idea of 'partnership' features strongly in the agreement. The term 'partnership' appears 52 times within the text of the Cotonou Agreement and a further nine times in the annexes (Raffer 2001, p. 3). The reality of the situation, however, is that despite their misgivings, the ACP countries have had Cotonou thrust upon them.

The post-Lomé partnership

As stated, and in line with the general development consensus, there has been a noticeable shift in the language relating to EU development policy over the past 20 years, the emphasis being placed squarely upon 'people-orientated', 'pro-poor' policies rather than economic goals and targets. The Treaty on European Union (TEU), signed in 1992, forms the basis for the current approach to development issues, with Article 130u stating that members 'shall foster the sustainable economic and social development of the developing countries, and more particularly the most disadvantaged among them'. Based on Article 130u, the European Commission (2000a, p. 5) has highlighted four key objectives where its development policy is concerned:

1. Reducing poverty;
2. Supporting economic growth by virtue of enhancing competitiveness in the rural sector;
3. Promoting good governance, democracy and human rights;
4. Improving environmental and institutional sustainability.

The sustainable development objective is further entrenched by Article VI of the Treaty of Amsterdam (1997), which requires that all EU policies, including those relating to development, be conscious of environmental considerations.

A further dimension has been added to EU policies through the adoption of the United Nations Millennium Declaration, unveiled in September 2000, which sets out targets for reducing poverty, gender inequality and environmental degradation. These objectives, together with other previously agreed upon International Development Goals, were subsequently amalgamated by the United Nations General Assembly in order to articulate eight 'Millennium Development Goals' (MDGs), (see Table 1.1). These targets now dictate the direction of EU development funding and it is against these MDGs that the success of EU policies must be judged (Cotonou Partnership Agreement 2000, Preamble). The Treaty of Nice (2003) demonstrates how the core elements have been incorporated into the EU agenda. Article 177 stresses that the EU should work towards 'the sustainable economic and social development of the developing countries, and more particularly the most disadvantaged among them, the smooth and gradual introduction of developing countries into the world economy, and the campaign against poverty in the developing countries'. Furthermore, Article 178 commits the EU to 'take account of the objectives referred to in Article 177 in the policies that it implements which are likely to affect developing countries'. Finally, Article 180 calls on both the EU and all the member states to 'coordinate their policies on development cooperation and ... consult each other on their aid programmes, including in international organisations and during international conferences'. Sustainable development and poverty eradication are thus firmly entrenched in the Treaties which constitute the legal basis for the EU. The objectives set out in the MDGs are most specific, and, if EU development policy is to be effective in aiding the realisation of these targets, then funding to developing countries must be carefully targeted so as to produce optimal returns.

While WTO rules do not allow for non-reciprocal agreements, they do allow for Economic Partnership Agreements (EPAs): agreements between economic blocs. It is for this reason that the EU chose to build the post-Lomé strategy on regionalism. Article 35 of the Cotonou Agreement stresses that 'economic and trade cooperation shall be built on regional integration initiatives of ACP states, bearing in mind that regional integration is a key instrument for the integration of ACP countries into the world economy'. The proponents of increased cooperation maintain that, in addition to attracting additional investment, regional integration will provide the ACP countries with economies of scale and increased competitiveness. Furthermore, regional integration is seen as an aid to overcoming the problems associated with being a small or landlocked state. The EPAs also have the potential to allow for broad, trans-boundary environmental standards. However, the Cotonou strategy represents a major departure from the Lomé regime and many ACP states felt, and continue to feel, that the timetable for ending non-reciprocal trade links was rushed

Table 1.1 The United Nations Millennium Development Goals (UN 2005)

Goal	Target
Eradicate extreme poverty and hunger.	Halve (1990–2015) the proportion of people living on less than $1 per day. Halve (1990–2015) the proportion of people suffering from hunger.
Establish universal primary education.	Ensure that by 2015 all children will have access to primary education.
Promote gender equality and the empowerment of women.	Remove gender disparities in education by 2015.
Reduce child mortality.	Reduce by two-thirds the under-five mortality rate by 2015.
Improve maternal health.	Reduce by three-quarters the maternal mortality ratio.
Fight HIV/AIDS, malaria and other diseases.	Begin to have reversed the spread of HIV/AIDS by 2015. Begin to have reversed the spread of malaria and other diseases by 2015.
Ensure environmental stability.	Integrate sustainable development principles into country programmes and policies and reverse the loss of environmental resources. Cut by half the proportion of people without access to safe drinking water by 2015. Improve the lives of at least 100 million slum dwellers by 2020.
Establish a global partnership for development.	Develop an open, rule-based, non-discriminatory trading system. Special consideration to be given to LDCs. Special consideration to be given to landlocked and small island states. The debt problems of developing countries to be dealt with. Strategies to secure productive work for youth to be introduced. Cooperate with pharmaceutical companies to provide affordable drugs to those in need of them. Facilitate transfers of technology to developing countries.

and could come to damage their economies. Consequently, at the outset of negotiations, the ACP countries campaigned vigorously for an extension to the WTO waiver relating to Lomé preferences. This extension was eventually secured, facilitating the gradual implementation of the Cotonou Agreement over a period of eight years, expiring on 31 December 2007. This waiver has allowed ACP countries a certain degree of 'breathing room' in order to prepare their economies for the loss of tariff revenues and other duties brought about by their accession to the new agreement.

However, there were, and remain, certain difficulties associated with the regional strategy put forward by the EU. Although, historically, there have been no shortage of regional groupings involving ACP members, by 2000 very few were in any sense 'advanced'. Examples of ACP regional groupings prior to Cotonou include:

- The Caribbean Community and Common Market (CARICOM)
- The Common Market for Eastern and Southern Africa (COMESA)
- The East African Community (EAC)
- The Economic Community of West African States (ECOWAS)
- The Pacific Island Forum (PIF)
- The Southern African Customs Union (SACU)
- The Southern African Development Community (SADC)
- The West African Economic and Monetary Union (WAEMU)

When one assesses the history of regionalism amongst the ACP states, the problems inherent in a regionally based development strategy become immediately apparent. Of the 53 countries in Africa, only six are members of just one regional organisation. Of the remainder, 26 states are members of two groupings, 20 simultaneously belong to three, whilst the Democratic Republic of Congo is a member of four. By 2000, only WAEMU, consisting of francophone West African states, was sufficiently integrated to contemplate the negotiation of a EU EPA. The WAEMU formed a customs union in January 2000 and, whilst successful in this regard, almost crippled ECOWAS by dividing the Anglophone from the Francophone states. The only other functioning African customs union, SACU, is in effect already part of an EPA by virtue of the free trade area established between South Africa and the EU (see pp. 27–8; p. 96). Regional integration in the other groupings mentioned was, and remains, limited.

Despite this limited progress in regional integration initiatives, the timetable set dictates that negotiations for the creation of EU-ACP EPAs, begun in 2002, must be completed by 2008, when the Doha-sanctioned waiver for the Lomé preferences expires. An additional complication for the ACP countries is that the ongoing WTO Doha Round negotiations are running parallel to the EPA talks. These parallel negotiations are problematic since, if the new EU-ACP agreement is to be WTO-compliant, talks should arguably be suspended until the new WTO rules are clarified. The Doha Round, originally timetabled for completion by January 2005, is well behind schedule (and, at the time of writing, set for completion by the end of 2007 – however, the Round may yet drag on until after the 2008 US Presidential elections) and could yet have important ramifications for any EU-ACP FTAs, especially given that paragraphs 28–29 of the Doha Ministerial Declaration, which relate to regional trading blocs, are points that still require clarification (see Chapter 9 for a detailed discussion of the EPA process).

There are also problems relating to funding. For the sake of comparison, it is useful to note the costs involved in the incorporation of the twelve 2004 accession states by the EU. The cost of integrating these new members will be an estimated €40 billion ($52 billion) (Gunessee 2002, p. 11). While the EU has contributed significant funds to the ACP countries for regional initiatives in the past, this funding is incomparable to the money on offer to the new EU members. The adjustment costs associated with the proposed EPAs are significant. They relate to the loss of tariff revenues, restructuring of domestic sectors, the development of the institutional capacity with which to deal with Sanitary and Phytosanitary (SPS) measures and the harmonisation of policies with neighbours, to highlight but a few. The Commonwealth Secretariat has estimated that adjustment costs will represent a minimum of €9 billion ($11.7 billion) to the ACP as a whole (Bilal and Rampa 2006). To put this figure into context, during the period 2001–04 the EU offered the ACP countries €300 million ($390 million) in trade-related assistance (Bilal *et al.* 2006a). The ACP countries have requested that the EU establish an adjustment fund in order to facilitate the transition to the new regime. However, the EU has stated that any adjustment costs will be covered by the tenth EDF which comes into effect in 2008 and has a commitment of €22.7 billion ($29.51 billion) (European Commission 2006b). There is little 'new money' in this aid budget and the fear is that, as a result, these adjustment costs will have to be met at the expense of other sectors such as health or education. Furthermore, there is a growing feeling of resentment in some quarters that EPAs are being forced upon the ACP countries. Edwin Laurent, chair of the ACP Ambassadors Trade Committee, is one such voice:

> I do not think that we should accept the inevitability of reciprocity and Free Trade Areas as if in fact this is what the Cotonou negotiations

> are about. If you look at the Cotonou Agreement, that commitment is not in there. If we do not negotiate successfully we could very well be faced with a situation in which the EU demands from us something of great value – essentially free access for European goods and services to our markets. But in exchange they give us nothing except the advice that if we are to develop we must open up our markets and liberalise. (Cited in Action for Southern Africa 2002b)

A mounting sense of self-confidence amongst developing countries over the past 15 years has led to a certain degree of opposition to the imposition of EPAs. A recent history of GATT/WTO negotiations demonstrates that developing countries can, if there is sufficient will, force compromises from developed countries. Before the Uruguay Round, the majority of developing countries saw little need to participate within these structures as they had little to gain from them. Export-substitution-based policies meant that developing countries were not focused on exports. Furthermore, preferential trade treaties such as Lomé meant that it was often more profitable for these countries to remain uninvolved in international trade negotiations. Consequently, developing countries had little input in the evolution of the GATT regime, which allowed developed countries to establish a framework that prioritised their needs: low tariffs on manufactured goods and high levels of protectionism for the agricultural sector. It was only with the onset of the Uruguay Round in 1986, by which stage developing countries were increasingly gearing their economies for export-led growth, that many chose to become more involved in negotiations. The conclusion drawn by most developing countries was that those countries that had been most involved in the bargaining, such as Brazil and India, also gained the most from the Round (Page 2002, pp. 4–5).

The Seattle Summit in 1999 demonstrated that collective action could be decisive, with developing countries conspiring to block any consensus at the meeting. The USA had planned to use the summit to push through a narrow trade liberalisation agenda. Developing countries, however, were determined that the WTO address agricultural issues as well as make a commitment to a 'Development Round' which would focus on poverty reduction. In order to attempt to salvage a declaration in the midst of serious disagreement the US hosts invited 20 delegates to partake in secret 'Green Room' talks in order to reach some level of consensus – no African countries were invited. This snub united developing countries in their opposition to the US-led agenda and as a result approximately one hundred members threatened to withdraw from the meeting (Madeley 2000, p. 21). Consequently, developing countries were able to block any agreement on a final declaration. Developing countries were bolstered in their resolve by the mass demonstrations that took place in Seattle, causing widespread disruption and focusing global attention on a number of contentious trade-related issues.

The full power of the developing bloc was even more amply demonstrated at the Fifth Ministerial Conference of the WTO, held in Cancun, Mexico, in September 2003. The talks collapsed with half a day to spare and without any consensus on the major issues timetabled for discussion. Developing countries were able to 'highjack' the summit as a result of their ability to form effective alliances, such as the G-20 Bloc which includes South Africa, Brazil and India. Poorer countries refused to bow to the proffered EU-USA agenda on offer at Cancun and instead demanded extensive reform of the agricultural sectors in the North. The EU complained about G-20 inflexibility and described Cancun as a 'missed opportunity' for developing countries. Franz Fischler (2003a), the member of the Commission then responsible for agriculture, rural development and fisheries, commented that

> there were too many divergent views, there was too little flexibility shown by too many people, and the structure of the WTO leaves too much space for inefficient decision making. It is also perhaps a peculiar twist of fate that the name of the city chosen for the latest ministerial meeting should mean snake pit in the local Mayan language.

The Americans were similarly unimpressed by this show of solidarity on the part of the developing countries. The US Trade Representative at the talks, Robert Zoellick, maintained that Cancun had divided WTO members into 'can-do' and 'won't-do' countries. He warned the 'won't-do' countries that the 'US will not wait [and] will move toward free trade with "can-do" countries' (cited in *Trade Negotiation Insights* 2004). The USA has subsequently succeeded in breaking a number of countries away from the G-20 group, including Colombia, Peru and Guatemala, by offering them lucrative bilateral trade arrangements. Both the USA and the EU can effectively bypass the WTO by virtue of free trade agreements and may yet choose this option if deadlock at the WTO continues. While Seattle, Cancun and, more recently, Hong Kong have demonstrated that developing countries are now taken more seriously within the WTO and, if sufficiently united, can determine the outcome of any further talks, the options for ACP countries outside of the current EPA negotiations are somewhat restricted. First, EU has managed to break the unity of the ACP bloc, ensuring that there would be no repeat of the 'ambush' of 1975. Furthermore, a refusal to consider the EPA strategy could effectively result in individual ACP countries being downgraded to the basic GSP for developing countries. Thus, while the bargaining power of developing countries in the WTO has progressed significantly since the early 1990s, their ability to sway EU-ACP negotiations is somewhat more limited in scope.

The 'Problem' of LDCs

Unlike 1975 and the signing of the Georgetown Agreement, obstacles to ACP unity abound. The Cotonou Agreement makes special concessions for Least Developed Countries (LDCs), thereby creating a two-tiered ACP group. Classification is important as, under the new agreement, LDCs are entitled to continued, non-reciprocal privileged access to EU markets – an arrangement sanctioned by the WTO and not necessitating a waiver. Article 85 of the Agreement states that: 'The Least Developed ACP states shall be accorded special treatment in order to enable them to overcome the serious economic and social difficulties hindering their development so as to step up their respective rates of development.'

To facilitate this development, ACP LDCs are currently party to the 'Everything But Arms' (EBA) initiative (the initiative will be discussed in some detail in Chapter 4), which allows for tariff-free access to EU markets for nearly all LDC goods. As a result of EBA, the EU will eliminate 919 tariff lines on LDC goods by 2009, with only 25 lines relating to the arms trade remaining. Tariff lines on sugar, rice and bananas were targeted to be removed gradually, so as to stagger the flow of these commodities into the EU. However, EBA creates certain problems for regional groups, with LDCs, under the terms of the Cotonou Agreement, having no real incentive to join the proposed EPAs. The EBA initiative allows LDCs tariff-free access to EU markets without the need for reciprocity. By joining a regional group and negotiating an EPA with the EU, LDCs stand to lose any tariff-derived income, with little chance of offsetting these losses. If LDCs prove reluctant to join regional EPAs, it could undermine the viability of a number of regional bodies, especially in Africa. The Commission has attempted to 'sweeten' the EPA preferences by offering tariff and quota free access to all ACP exports to the EU (ICTSD 2007). However, a number of EU member states have balked at this proposal.

The SA-EU FTA: a blueprint for negotiations?

The negotiations that led to the South Africa–EU free trade area have raised a number of issues for the ACP countries. As mentioned above, these negotiations were prolonged and often acrimonious. The result has been a lingering suspicion that the EU is not above bullying weaker countries into accepting its policies. Negotiations between South Africa and the EU for a Trade, Cooperation and Development Agreement (TCDA) were initiated in June 1995 and concluded in March 1999. The TCDA, which sets out the creation of a FTA and also outlines future EU financial assistance to South Africa in the form of the European Programme for Reconstruction and Development (EPRD), was only concluded after 24 rounds of formal talks as well as numerous informal sessions.

The final agreement was a far cry from that which the post-*apartheid* South African government had originally envisaged, centred as this had been on South Africa becoming a member of the ACP bloc and acceding to the Lomé Convention. The EU denied this request on the grounds that it viewed South Africa as a developed country and that such a step would not be WTO-compatible. Furthermore, it soon became evident that the EU wished to use any agreement with South Africa as the basis for future agreements with other ACP countries (Lee 2002, p. 86). It was argued that, as a developed country, for South Africa to have improved access to EU markets would necessitate the creation of a FTA. Concerned about the prospect of a FTA, the South African government rejected the EU's proposals. As a result, negotiations broke down and were only resumed in January 1997. In April 1997, as a concession, South Africa was allowed to accede to Lomé, although not as a full member. This step paved the way for further negotiations and, consequently, by October 1999, the TDCA had been signed by both parties. Yet a number of problems remained unresolved. Wine and spirits became the focal point of sometimes bitter talks. A number of EU member states objected to South Africa making use of the terms 'port' and 'sherry', claiming that this practise would undermine European producers. Further objections were raised against the use of 'ruby', 'ouzo', 'grappa' and 'grand cru'. After a tele-conference in June 2000, it was believed that an agreement had been reached. However, on perusal of the proffered EU text, South African negotiators argued that the EU had attempted to 'slip' new clauses into the agreement and on the basis of this refused to sign it. Commissioner Franz Fischler urged South Africa to sign the agreement and to negotiate changes once the agreement had come into force. A statement by the South African Department of Trade and Industry highlighted the growing tensions:

> The changes made by the EU after the 9th June are not acceptable to us as they are matters of substance and cannot be resolved technically at a later stage. Once again the EU seeks to reduce South Africa's multilateral rights and extend the names it wishes to protect.... This is unfortunate and it is sad that we repeat again the pressure negotiating tactics. (Cited in Action for Southern Africa 2002a)

A compromise on the matter was eventually reached towards the end of 2001 when the EU agreed to increase South Africa's duty-free quota for wine and spirits from 32 million litres to 42 million litres per annum in return for South Africa refraining from the use of the contested descriptions (Action for Southern Africa 2001b).

To many developing countries, these negotiations fostered the perception that the EU was not prepared to provide poorer countries with meaningful concessions. This sense was further underlined by EU demands that EU fishing interests be allowed access to South African waters – a demand largely viewed by the South African delegation as unreasonable. The often adversarial nature of the EU-South African negotiations has been interpreted by many ACP countries as being an indication of the EU's unwillingness to be flexible when considering development policy. The negative perception engendered by the EU-SA negotiations has prevailed and has, to an extent, poisoned the atmosphere surrounding the talks. A 2007 Tearfund/Tradecraft paper clearly articulates this view of the Commission as a 'bully', outlining the one-sided nature of the EPA negotiations (Griffith and Powell 2007). The overall result has been a marked lack of enthusiasm for Cotonou amongst the ACP countries, with a Ghanaian trade official declaring somewhat morosely that 'we will not survive in this game' (EPA2007.org).

Conclusion

The 'special' relationship between the EU and the ACP, although still trumpeted, has slowly eroded over the past three decades. In its initial incarnation, Lomé offered developing countries unprecedented access to developed country markets. However, this special relationship has been curtailed for a number of reasons: the Washington Consensus, the end of the Cold War, the evolution of the international trading regime and the rise of the EU as a genuinely international actor. The consequence of these factors for the ACP countries is that EU relations with the Bloc have become 'normalised'. There is an increasing sense that the ACP countries are no longer special, but are now simply one trading bloc amongst many. As a result, when Lomé expired in 2000 it was replaced with an agreement that differed little from EU trade agreements with other developing countries. The Treaty is firmly locked into the neoliberal framework that has been evolving amongst key international donors since the 1980s and, unlike the early incarnations of Lomé, Cotonou deviates little from the internationally accepted 'line'. The following chapters evaluate the likelihood of this regime engendering poverty alleviation, sustainable development and the successful integration of the ACP countries into the global economy. 'Sustainable development' arguably lies at the core of this assessment, as any progress made by the ACP must, out of necessity, be sustainable in the long term. However, the term is one that is contested and it is important to outline the nature of what the EU's interpretation of sustainable development entails as this, according to the Agreement, underpins the Polity's approach to development.

2
Towards an EU Definition of Sustainable Development

Sustainable development has purportedly been a guiding strategy for all EU policymakers since the signing of the Treaty on European Union (TEU) in 1992, affecting both domestic and foreign policy requirements, including those concerning the Polity's relations with developing countries (Article 130u): 'The Commission undertakes in its proposals, and the member states undertake in implementing these proposals, to take full account of their environmental impact and of the principle of sustainable growth.'

This commitment was made more explicit in the Treaty of Nice (2003 – Article 175):

> The High Contracting Parties are determined to see the European Union play a leading role in promoting environmental protection in the Union and in international efforts pursuing the same objective at a global level. Full use should be made of all possibilities offered by the Treaty ... to promote sustainable development.

Sustainable development entails more than environmental protection, encompassing as it does a far wider array of issues. As an approach, therefore, it comprises elements above and beyond simple conservation strategies, including social equity and economic development. It is thus not always appropriate to employ the terms 'sustainable development' and 'environmental protection' within the same context. However, attitudes towards environmental policy and its implementation form a crucial component of this widely utilised umbrella term. The EU's interpretation of sustainable development is significant; it is this definition that provides the measure for assessing the success or failure of any given EU development initiative. It is therefore important to set out the debate surrounding sustainable development, and to contextualise the EU's interpretation, and the potential ramifications thereof, within this wider debate.

Sustainable development

The origins of the sustainable development paradigm arguably date back to 1972 and the UN Conference on the Human Environment (the Stockholm Conference). Although this was by no means the first time the international community had convened to discuss environmental concerns, with Stockholm being preceded by international conferences relating to whaling, fisheries and pollution, this was the first international attempt to discuss the environment in a comprehensive and systematic manner. Many of the key debates outlined in 1972 remain pertinent, including the differing roles and responsibilities of both the North and the South in preventing environmental degradation. One of the most contentious notions permeating the Conference was the issue of possible limits to economic growth. The *Limits to Growth* thesis, published shortly before the Conference by the Club of Rome,[1] involved projections pertaining to the sustainability of economic growth. The findings suggested that the then current rates of growth would result in the depletion of natural stocks and, ultimately, in environmental catastrophe. The computer models employed indicated that, even with technological advances, eventual environmental collapse was inevitable. In addition, the computer models prompted the conclusion that environmental concerns could not be compartmentalised and dealt with in isolation from economic factors; governments needed to develop strong, overarching environmental policies. The thesis duly triggered a significant degree of debate, particularly after the pessimistic baseline assumptions underlying the computer models were called into question (Dobson 1995, pp. 72–6).

The South reacted with scepticism when the *Limits to Growth* hypothesis was proffered. There was a perception amongst many of the delegates from developing countries that the North was attempting to use environmental concerns to block the future development of the South. Having ruined their own environments, developed countries were, in this view, endeavouring to prevent developing countries from doing the same. Developing countries argued strongly that a lack of economic growth was also environmentally destructive. The argument was that any global response to environmental problems would, as a prerequisite, have to include development considerations (Conca and Dabelko 2004, pp. 18–19). Indira Gandhi articulated this view most forcefully, 'of all the pollutants we face, the worst is poverty. We want more development' (cited in George and Kirkpatrick 2003, p. 122). This view was echoed in a 1972 article by the Brazilian Ambassador to the US, João Augusto de Araujo Castro, in which he argued:

> There is a pollution of affluence and a pollution of poverty. It is imperative to distinguish between the two lest some pollution be prevented at the cost of much economic development. Were it not for the dangers arising from the confusion between the two kinds of pollution, there would be no need for calling attention to the precarious housing conditions, poor health, and low sanitary standards not to mention starvation in developing countries. The linear transposition of ecological problems of the developed countries to the context of the developing ones disregards the existence of such distressing social conditions. Wherever these conditions prevail, the assertion that less income means less pollution is nonsense. It is obvious, or should be, that the so-called pollution of poverty can only be corrected through higher incomes, or more precisely, through economic development. (p. 409)

Much of this debate was, however, confined to academic and political circles. It was arguably only in 1987, with the publication of the Bruntland Report, that the sustainability debate truly entered the political and public arena. The Report highlighted the links between environmental and economic policy and effectively ushered in the era of sustainable development. It defined sustainable development as 'development that meets the needs of the present without compromising the ability of future generations to meet their own needs' (World Commission on Environment and Development 1987, p. 43). What the Report implied was that a radical change of institutions was not the sole means of preventing further degradation. It maintained that economic and environmental policy could be reconciled and that dramatic action to stave off environmental collapse was not obligatory. In this form, sustainable development can be viewed as an attempt to couple economics with environmentalism so as to demonstrate that the two are not necessarily incompatible. While there was, and remains, general consensus as to this basic starting point, the proverbial devil is in the details – how to combine capitalist economic practices with environmental protection. Historically, capitalism has proved inept at both producing public goods and in taking a long-term view, and it is for this reason that sustainable development incorporates a number of contradictions.

Before the advent of the concept of sustainable development, economists tended to view economic growth as a closed cycle connecting the different sectors of the economy. Environmental factors were generally ignored. This occurred because neoclassical economic theory has always assumed a system operating a long way short of its limits. Any environmental degradation in such a system would be largely irrelevant. Current indications suggest that the system might be operating closer to its limits than generally anticipated, thus pointing to a need for environmental considerations to be factored into the equation. Depletion of natural resources, pollution and the threat to life support mechanisms can all be said to be economic by-products and must thus be accounted for (Jacobs 1991, p. 5). The 'strong' versus 'weak' sustainability debate centres on the extent to which it is believed that these issues should be addressed.

Strong versus weak sustainability

Sustainable development is regularly put forward as the solution to high levels of environmental degradation. As an economic strategy it has been adopted and promoted by a diverse range of interests which ordinarily would have little in common. Environmentalists, business interests, government institutions and donor organisations have all expressed their approval of sustainable development as a viable strategy for combating environmental degradation. This broad-based appeal stems largely from a limited consensus as to what sustainable development actually entails. There is no single, accepted definition of the term. Proponents of sustainable development can, however, be roughly divided into two camps; advocates of either 'strong' or 'weak' sustainability. The divide is significant because the two sides offer dramatically differing solutions to the question of environmental protection. This has implications for policymakers who purport to advocate sustainable development, offering as it does alternative strategies covered by a single 'catch phrase'. However the weak/strong divide is a contentious one. While extreme positions are readily defined and compartmentalised, what remains is somewhat more problematic and is therefore discussed below in some detail.

Weak sustainability

While sustainable development entails more than simply addressing conservation issues, attitudes towards environmental degradation highlight particular 'shades' within the wider debate. In its most basic form, weak sustainability involves the protection of total capital stock (constant capital rule) and presupposes that all forms of capital are substitutable for each other. Degradation of one form of capital must be balanced by ensuring capital growth in other areas. However, for the purposes of this study, the definition of weak sustainability will be broadened somewhat to incorporate strategies that do not, in essence, challenge the prevailing neoliberal economic paradigm. Such strategies argue that basic reforms to the current system (for example, improved accounting techniques) can help stem the tide of environmental degradation without dramatically altering the economic *status quo*. In fact, it can be argued that such steps would make the global economy more efficient. The weak position, then, is essentially one of minimums: we must do whatever is necessary to prevent environmental catastrophe while simultaneously allowing for the right of future generations to enjoy at least our own present standard of living (Jacobs 1991, p. 72).

Much of the debate surrounding attempts at defining weak sustainability centre on the conservation of natural resources, or 'natural capital',[2] as it is often described. As stated, orthodox economic theory argues in favour of the preservation of a constant stock of 'total' capital, namely human-made and natural capital. The implication is that natural capital can be substituted with human capital and that human ingenuity will be able to counterbalance any negative effects brought about by degradation. This can be interpreted as a 'narrow' view of sustainability, whereby a determined level of 'total' stock is passed down from one generation to the next. This view has recently been afforded an increased measure of credibility as a result of the enhanced public profiles of such environmental sceptics as Bjørn Lomborg and Matthew Ridley.

However, it was Julian Simon who pioneered this anti-environmental backlash. Writing in the early 1980s, he produced a number of influential rebuttals to the rising tide of environmental doomsayers such as Paul Ehrlich, author of *The Population Bomb* (1968), a neo-Malthusian work. Simon and Ehrlich's famous $1000 wager on the future of metal prices proved a damaging episode for the neo-Malthusian lobby and earned Simon the epithet of 'Doomslayer'.[3] Simon argued that observable trends suggested that the quality of human life on earth was becoming better rather than worse, with warnings of environmental catastrophe having been overstated or based on unsound science (Simon and Wildavsky 1984, pp. 171–83). In 1984, Simon, together with Herman Kahn, published a response to the US government publication *Global 2000: A Report to the President*, a widely disseminated study of environmental degradation commissioned by President Jimmy Carter in the late 1970s. The report, written before the Reagan years of ascendant 'new right' economic theory, suggested that by 2000 the world would be overcrowded, heavily polluted and ecologically unstable. It went on to emphasise how 'if present trends continue... life for most people on earth will be more precarious than it is now' (cited in Simon and Kahn 1984, p. 1). Simon and Kahn (1984, p. 45), however, argued that the available evidence suggested the opposite:

> Our conclusions are reassuring, though not grounds for complacency. Global problems due to physical conditions (as distinct from those caused by institutional and political conditions) are always possible, but are likely to be less pressing in the future than in the past. Environmental resource, and population stresses are diminishing, and with the passage of time will have less influence than now upon the quality of human life on our planet. These stresses have in the past always caused many people to suffer from lack of food, shelter, health and jobs, but the trend is toward less rather than more of such suffering. Especially important is the trend toward longer and healthier life throughout the world. Because of increases in knowledge, the earth's 'carrying capacity' has been increasing through the decades and centuries and millennia to such an extent that the term 'carrying capacity' has by now no useful meaning. These trends strongly suggest a progressive improvement and enrichment of the earth's natural resource base, and of mankind's lot on earth.

Simon's influence has been widespread and his ideas have steadily been afforded a much wider audience. Matthew Ridley's long-running column in *The Sunday Telegraph*, entitled 'Down to Earth', later published in collected form in 1995 and 1996 by the Institute of Economic Affairs (IEA) (a free-market think-tank founded in 1955), served as a forum from which he was able to rail against environmental legislation, cast aspersions on the evidence for climate change and generally prescribe a loosening of government control in environmental matters. These columns struck a cord with a sizeable proportion of the British public, together with sectors of the business community. The foreword to the first collected volume, by John Blundell, the General Director of the IEA, describes Ridley as being:

> [i]n the vanguard of one of the greatest intellectual battles of our time, a battle which will eventually determine the kind of society in which we live. On the surface it is a conflict between two visions of the environment: whether it is best managed by private or public means... It is an old battle dressed up in 'green' clothes. (Blundell 1995, p. 5)

However, it is Lomborg's tome, *The Skeptical Environmentalist*, which has succeeded in bringing the debate most sharply into focus. The book has polarised opinion on both sides of the Atlantic, being feted by publications such as *The Washington Post* and *The New York Times*, and pilloried by environmental organisations such as the Green Alliance. In Britain the launch of the book saw Lomborg being called upon to contribute articles to publications as disparate as *The Economist* and *The Guardian*.

Lomborg's position is one very much based on that of Simon; he argues that, far from being on the brink of environmental collapse, conditions are improving. He maintains that environmental 'doomsayers', the mass media and organisations such as Greenpeace, Friends of the Earth and the World Wildlife Fund for Nature, have grossly exaggerated the extent of environmental degradation and have indoctrinated the general public with their overly pessimistic views. Lomborg stresses that levels of species extinction and biodiversity loss are in reality a fraction of what has been claimed. Furthermore, contrary to the claims of environmental groups, global forest cover is increasing rather than falling, landfill wastes are not threatening to overwhelm the planet and the issue of global warming, even if it exists, is unlikely to be resolved by 'absurdly expensive' measures such as the Kyoto Protocol:

> Global warming, though its size and future projections are rather unrealistically pessimistic, is almost certainly taking place, but the typical cure of early and radical fossil fuel cutbacks is way worse than the original affliction, and moreover its total impact will not pose a devastating problem for our future. Nor will we lose 25–50 percent of all species in our lifetime – in fact we are losing probably 0.7 percent. Acid rain does not kill the forests, and the air and water around us are becoming less and less polluted. (Lomborg 2000, p. 4)

Lomborg (2001, pp. 32–3) argues that the line between good intention and truth has become blurred where the environmental debate is concerned, resulting in the stultification of any real discussion of the issues. Instead of questioning the perceived state of the environment, the public has been misled by half-truths, questionable science and statistical manipulation. Lomborg maintains that this indoctrination is not acknowledged because dissenters are dismissed as right-wing apologists. He attempts to reassure his readers as to the quality of his own research by stressing the fact that *The Skeptical Environmentalist* has approximately 3000 references and is thus a thoroughly researched critique of environmental orthodoxy.

Lomborg posits the view that 'environmental development often stems from economic development – only when we get sufficiently rich can we afford the relative luxury of caring about the environment' (2001, p. 33). Accordingly, he provides statistics that suggest that there is a correlation between levels of GDP per capita and environmental sustainability. By implementing 'expensive' policies such as the Kyoto Protocol, environmental sustainability is actually being put at risk by the potential slowing of economic growth. Lomborg and Simon both argue that human ingenuity is capable of dealing with almost any environmental threat, stressing that throughout history humans have overcome shortages and constraints by virtue of substitution and technology, and through being adaptable to circumstances (Kysar 2003, pp. 223–80).

Such arguments have galvanised opposition to the 'optimists' in general and Lomborg in particular. Some of this opposition has been more vociferous than others. He has been derided by his own government (Denmark), which accused him of 'scientific dishonesty' (Danish Committee on Scientific Dishonesty 2002, p. 9). The Danish Committee on Scientific Dishonesty stated that 'there has been such perversion of the scientific message in the form of systematically biased representation that the objective criteria for upholding scientific dishonesty have been met' (cited in *The Guardian*, 9 January 2003). These accusations have since been withdrawn. However, this has not served to calm the ferocity of the debate. Lomborg has been accused of selectively reading data, demonstrating a lack of scientific expertise and of exaggerating certain claims made by the environmental movement. For example, he regularly cites Isaac Asimov, a science fiction author, as a key spokesperson for the environmental movement; a position not afforded Asimov by the majority of environmentalists. Furthermore, Lomborg's much-vaunted 3000 references are somewhat 'thin' when subjected to closer scrutiny. He makes sweeping generalisations regarding the claims of the environmental movement' based on scanty reports from the popular press and little-known publications rather than from information sourced from manifestos or official press releases. He then proceeds to attack these 'straw men' with relative ease (Burke 2001, pp. 4–5). Critics have also stressed that despite Lomborg's misgivings relating to the accuracy of computer modelling of weather patterns by analysts studying global warming, he is content to accept economic computer modelling as accurate despite both forms of modelling being subject to similar constraints. This willingness to accept the implications of economic modelling without question has led certain commentators to rechristen his book *The Gullible Economist* (Lowe 2003).

Calmer commentators, such as Eric Neumayer (2001), have praised Lomborg's attempt to engage with the debate whilst criticising his failure to contribute anything that is substantially new to the discussion, other than a reiteration of the basic optimist viewpoint. He also argues that Lomborg's central thrust against the proponents of the 'litany', that many of their claims are exaggerated, is somewhat irrelevant. First, the rate of species extinction, even if it is as low as Lomborg claims, is still unacceptable to many people and, second, even if groups such as Greenpeace and WWF have exaggerated various claims, these groups have succeeded in rousing the public and policymakers to act in favour of environmental protection. He notes that it is precisely because of these actions that the declining pollution levels in developed countries have been achieved.

For Lomborg's detractors, market forces, substitution and an ability to adapt to changing circumstances will not be enough to solve or negate all environmental problems. Many critics stress that in terms of life support function, certain aspects of nature are irreplaceable and that facets of the environment are unique and cannot be replaced with human-made substitutions. While human capital can be created through knowledge and can generally be recreated if destroyed, the same cannot be said for all natural resources (Jacobs 1991, pp. 82–3). Irreversibility and uncertainty are important factors to consider when assessing the optimists' view of sustainability. Opponents point to the ACP island state of Nauru in order to demonstrate the inherent dangers associated with the perceived substitutability of natural capital. In 1900, rich phosphate deposits were located on the island. These deposits have been extensively mined over the past century with the result that the island is now almost completely barren (McDaniel and Gowdy 2000, pp. 179–80). In exchange for the destruction of their environment, Nauruans were well compensated with a government-controlled trust fund worth approximately Australian $1 billion (US$ 770 million) by the 1980s (McDaniel and Gowdy 2000, p. 49). Islanders were to live off the interest once all phosphate stocks were depleted. However, poor management combined with the 1997 Asian financial crisis severely depleted the island's coffers and the islanders face a relatively uncertain future. The most significant problem associated with equating human and natural capital is that it presupposes that the two can be calculated in the same way. This is difficult, partly because natural resources perform a multitude of functions over and above their perceived market value (Jacobs 1991, p. 83). Consequently, it is difficult to arrive at a true 'total' value of natural capital.

It is becoming an increasingly accepted view in economic circles that a significant degree of environmental degradation is the result of certain market failures. Environmental economists argue that the reason environmental degradation occurs is because, as a rule, conventional economic theory factors are neither the real cost of natural resources subsumed nor the cost of the pollution generated in the production process. Consequently, final prices rarely reflect the true cost of production. Accordingly, adjustments in accounting procedures to incorporate these costs into the final price of the product would correct such market failures and thereby significantly diminish degradation. Therefore, it can be argued, if current market failures can be corrected, then high levels of environmental degradation need not occur (Barry 1999, pp. 145–7). However, the problem of assigning prices makes this position fraught. How are costs assigned and how are they then incorporated into the final price? Such calculations are important, since optimal pricing structures can only be achieved if cost estimates border on certainty. If cost estimates are inaccurate then continued market failure is a likely outcome (Panayotou 1993, pp. 38–40).

David Pearce, a leading exponent of neoliberal environmental economics,[4] stresses the need to maintain what he terms 'levels of critical natural capital'. These are levels of natural resources that are essential for the prevention of environmental catastrophe and the preservation of life support systems. The natural environment serves as both resource supplier and waste assimilator, and if these functions are to be retained, then the rate of use must be less than or equal to the rate of consumption (Pearce and Turner 1990, p. 44). Only if human capital is capable of replicating all the functions of natural capital can it be deemed substitutable. In the case of life support systems, it cannot. Consequently, Pearce (1992, p. 392) advocates keeping levels of natural capital constant, not in terms of maintaining specific levels of natural resources but in conservation as a whole. This means that it would be possible to have fluctuating levels of specific resources as long as the overall stock of natural capital does not decline. For Pearce, the key to sustainable development is that market failures be corrected, that 'harvest does not exceed regeneration' and that the economy does not outgrow its carrying capacity. The latter, he asserts, can be achieved through technological advances and by emphasising a switch from non-renewable to renewable resources (Pearce *et al.* 1994, p. 66).

Pearce (1993, p. 15) is anxious to demonstrate that the mutual suspicion that exists between environmentalists and economists is misplaced. According to his argument, what is needed is a clear demonstration that the sustainable use of resources can have positive economic value: 'If ... conservation and the sustainable use of resources can be shown to be of economic value, then the dialogue of developer and conservationist may be viewed differently, not necessarily of opposites, but of potential complements.'

An optimally functioning market is the best way to ensure the efficient allocation of resources. In order for this to occur, all costs must be incorporated into the final price of a commodity. Invariably this creates difficulties, as no market exists for environmental services such as 'carbon sinks', and thus values can be difficult to assign. Pearce (1993, p. 5) argues that although 'valuation may be imperfect ... some valuation is better than none'. It is only by assigning an economic value that environmental costs can be ascertained. Market failure occurs when the price of a commodity does not fully reflect the costs inherent in its production. Market failure is particularly marked when costs are distributed widely throughout society. If these costs are not internalised, they can give rise to a misallocation of resources as a result of incorrect market signals, thereby increasing environmental degradation. Tradable pollution permits and 'green' taxes are strategies which can be employed in order to 'right' the imbalances created by inefficient accounting procedures.

Moreover, Pearce emphasises the need to expose misconceptions relating to the costs associated with environmental regulation. Macroeconomic models centring on the costs connected with environmental protection were developed during the early 1990s in Britain, Norway and the Netherlands and for the EU as a whole. The results were largely counter intuitive. In the majority of cases, the costs associated with environmental protection, such as carbon taxes, had little or no effect on GNP and in many instances actually served to boost employment figures through the development of pollution abatement industries (Jacobs 1991; Pearce 1993, pp. 95–104). Similar studies were conducted in the USA. However, these results were less clear-cut than in the European sample. Modelling suggested that the cost of implementing environmental legislation in the USA was twice that in Europe (Pearce 1993, p. 102). Pearce suggests that, despite the 'anomalies' presented by the American studies, environmental legislation is clearly not a hindrance to economic growth. If environmental regulation does not hamper economic growth, then there is no reason for economists and the business community as a whole to treat it with such suspicion.

More recent studies on the introduction of carbon taxes have produced mixed results. The potential impact of carbon taxes on growth depends largely upon the models employed, the baseline assumptions used and what the revenue generated by the proposed tax is to be spent on. It has been found that the use of either 'pessimistic' or 'optimistic' baseline assumptions when modelling the impact of carbon taxes on growth essentially predetermines the outcome. Such assumptions are based on factors such as product substitution, technological response, availability of alternative fuels and international cooperation. If the baseline assumptions are optimistic, then the models suggest little impact on growth (Cuervo and Gandhi 1998, p. 4). Some studies have actually shown a positive impact on GDP.[5] However, where pessimistic assumptions are employed, models suggest that carbon taxes can play a significant role in slowing economic growth.

Yet, despite discrepancies amongst proponents, the basic appeal of weak sustainability, as it is outlined above, stems from its compatibility with the current economic paradigm; economic and environmental policy need not be at odds with each other. It operates comfortably within a neoliberal economic framework and therefore offers no significant threat to the existing economic order. However, it can also be argued that weak sustainability offers a solution to environmental degradation without threatening the system largely responsible for the problem. In essence, it is a strategy by which you can have your cake and eat it too.

Strong sustainability

For some critics, the solution to environmental degradation is less straightforward than the proponents of weak sustainability might suggest. This is because, it can be argued, the current environmental crisis is also an economic crisis. (Jacobs 1991, p. 11). Degradation occurs as a result of economic activity and economic activity is threatened by the deterioration of the environment. Pollution and degradation have traditionally been viewed as unforeseeable, accidental side-effects of industrial development. However, while the effects of industrial development on the economy might be unintended, the results are certainly foreseeable. Michael Jacobs (1991, p. 25) suggests that while the 'invisible hand' of the market is readily acknowledged, the destructive, unintended 'invisible elbow' that follows in its wake is usually not.

In some quarters there is a general perception, based on the above, that attempts to reform the current economic model will lack the robustness necessary to arrest environmental degradation. Herman Daly (1993, p. 268), a leading advocate of strong sustainability, argues forcefully that there are limits to growth and that to 'delude ourselves into thinking that growth is still possible if only we label it "sustainable" or colour it "green," will just delay the inevitable transition and make it more painful'.

For Daly, finitude,[6] entropy[7] and complex biological interdependence all place limits on the economic system. There are limits to growth because, in observance of the second law of thermodynamics,[8] natural capital is finite and once it has been converted into some form of human capital, its energy is no longer economically exploitable and it eventually returns to the environment as waste (Daly 1992, p. 226). Furthermore, complex biological interdependence implies that, as the natural world is not fully understood, it is impossible to judge accurately what the outcome of tampering with it will be. Daly argues that the reason that the full extent of the problem is not acknowledged by economists is that those economists who factor environmental considerations into their calculations tend to do so at the microeconomic level, 'getting prices right' through the judicious use of cost–benefit analysis. The same economists also make much of optimal allocation and resource use. Daly (1996, p. 48) maintains that, by concentrating almost exclusively on the microeconomic level, economists fail to grasp the fact that, just as micro-units are a subsystem of the macroeconomy, so too is the macroeconomy a subsystem of the natural world and that within this environmental macroeconomy there are limits. Daly likens carrying capacity to that of the Plimsoll line on a ship. Optimal allocation of resources ensures that the ship can be more effectively packed. However, once the watermark touches the Plimsoll line the ship is full. Even if additional cargo is optimally positioned, the ship will still sink. It is important therefore, Daly argues, to concern ourselves not only with optimal allocation but also with optimum scale. Markets are highly effective tools for solving problems of allocation and provide both information and incentive. However, markets are incapable of addressing questions of distribution and scale (Daly 1996, p. 50). Thus, to rely upon market-driven strategies to provide sustainable development is to ignore these fundamental weaknesses. Daly argues that the notion of scale has already been inadvertently accepted by many economists and points to proposals relating to tradable greenhouse gas permits. The designers of these schemes accept in advance the necessity of a set limit to emissions, with trading then taking place within these boundaries. Such schemes are often touted as examples of market-driven strategies. This is undoubtedly true; they are market-based solutions but only within fixed parameters. Likewise, Daly maintains, it is necessary to try to determine the optimal scale for the global economy as a whole: a global 'Plimsoll line'.

Consequently, Daly (1992, p. 249) argues that the concept of growth in terms of GDP must be abandoned in favour of what he describes as 'steady-state economics', essentially a policy of strong sustainability:

> What is being sustained in Sustainable Development is a level, not a rate of growth, of physical resource use. What is being developed is the qualitative capacity to convert that constant level of physical resource use into improved services for satisfying human wants.

Daly defines 'growth' as a quantitative increase in the scale of the physical dimensions of the economy while 'development' is described as a qualitative improvement in the structure, design and composition of physical stocks and flows. There may be limits to growth but not necessarily limits to development. Only by switching the focus to development, rather than growth, can sustainable development become possible. Daly (1996, p. 3) argues that

> while sustainability is certainly relevant to the South, the critical issue is for the North to attain sustainability in the sense of a level of resource use that is both sufficient for a good life for its population and within the carrying capacity of the environment if generalised to the whole world. Population growth and production growth must not push us beyond the sustainable environmental capacities of resource generation and waste absorption. Therefore, once that point is reached, production... should be for replacement only. Physical growth should cease, while qualitative improvement continues.

This vision of development is not a new one, a point readily acknowledged by Daly; a similar strategy was proposed by John Stuart Mill in his *Principles of Political Economy* (1848), in which he advocated the need to embrace the 'stationary state', a condition which would involve no additional growth in population or physical capital stock but rather a refinement of technology and ethics. Mill (1994, p. 129) argued that

> [i]t is scarcely necessary to remark that a stationary condition of capital and population implies no stationary state of human improvement. There would be as much scope as ever for all kinds of mental culture, and moral and social progress; as much room for improving the Art of Living and much more likelihood of its being improved, when minds cease to be engrossed by the art of getting on.

Mill was, in essence, arguing in favour of development without growth. Such a view was dramatically at odds with the general view at the time that maintained that an end to growth would signal an end to profit and thereby an end to any motivation for further development. Wages would then stagnate, signalling an end to human 'progress'.

However, Daly has been heavily criticised, even by those favouring strong sustainability, for his anti-growth stance. Michael Jacobs (1991) counters a number of Daly's arguments: economic growth is not necessarily bad for the environment; it depends on which sectors of the economy are growing. A growth in teaching services or paper recycling will not impact negatively on the environment and may even have positive results. Similarly, a rise in GDP does not necessarily equate with a rise in levels of degradation. However, Jacobs does agree that there are limits to resource use. The destruction of natural capital cannot be viewed as simply a technical problem to be overcome by human ingenuity, as these measures cannot be assured. Jacobs stresses the need to include the environment in calculations of GDP. Degradation must be measured just as closely as growth. For Jacobs, if there is to be sustainable development then a number of factors are crucial. Not only do environmental concerns need to be an integral part of economic policy but also a fair distribution of wealth and certain levels of conservation are also imperative. This is because 'development' entails more than just growth. It has a welfare component as well as other non-financial concerns. Within this welfare component lies the issue of North–South relations. If sustainable development is to be possible, then these inequalities need to be addressed. Such problems cannot be solved through current economic strategies and call for such intervention as will shift the priority of economic activity from growth in an aggregate or global sense to one of targeting specific areas for growth.

For some, however, the circle cannot be squared. Mark Sagoff (1998, pp. 37–9), a colleague of Daly, maintains that too much effort has been expended in trying to justify strong sustainability in terms of 'carrying capacity' or 'limits to growth'. He discounts a number of Daly's assertions and argues that there need not necessarily be any limits to economic growth. Sagoff claims that Daly's understanding of growth and GDP is flawed. He argues that increases in GDP relate to the increased value of what is being produced rather than the actual physical size of the economy. Consequently, if the physical size of the economy increases but its value decreases, then it is actually experiencing negative growth. As a result, Sagoff contends that Daly's attempt to find an economic rationale for strong sustainability is unworkable. Renewable energy, for instance, could eventually negate entropy concerns. Rather than focusing on limits to growth as a 'call to arms', Sagoff suggests that this approach be abandoned in favour of a non-utilitarian conception of value. He cites the writings of John Muir, who railed against the 'temple destroyers' who sought only profit from nature, and Henry David Thoreau, who revelled in his relationship with the land, as more effective templates for justifying the need for environmental protection (Sagoff 1998, pp. 28, 45).

Conclusion

The differences between strong and weak sustainability are important, offering two dramatically differing options for governments and bodies such as the EU. The weak position suggests that reforms to the system and better accounting will allow for adequate levels of sustainability, whilst the strong position demands more active involvement from political institutions and a dramatic change in the global economic system. It is also important, when assessing the relative success/failure of the EU to ensure that Cotonou does in fact promote sustainable development, to have a 'yardstick' by which to evaluate the spirit of the EU-ACP partnership. The EU defines sustainable development in much the same way as the Bruntland Commission:

> Sustainable development means that the needs of the present generation should be met without compromising the ability of future generations to meet their own needs. It is an overarching objective of the European Union ... governing all the Union's policies and activities. It is about safeguarding the earth's capacity to support life in all its diversity and is based on the principles of democracy, gender equality, solidarity, the rule of law and respect for fundamental rights, including freedom and equal opportunities for all. It aims at the continuous improvement of the quality of life and well-being on

> Earth for present and future generations. To that end it promotes a dynamic economy with full employment and a high level of education, health protection, social and territorial cohesion and environmental protection in a peaceful and secure world, respecting cultural diversity. (European Council 2005, p. 28)

Further analysis of EU policy clearly demonstrates the polity's adoption of aspects of the weak position and its heavy reliance on strategies put forward by environmental economists such as Pearce, indicating its desire to retain much of the neoliberal economic *status quo*. This market-friendly approach is clearly outlined in the EU's Sixth Environmental Action Programme entitled 'Environment 2010: Our Future, Our Choice':

> To date, the approach towards business has largely revolved around setting standards and targets and then ensuring companies comply with these standards. Member States have increasingly supplemented this with market-based instruments, such as environmental taxes on different products, which aim to change the price signals in the marketplace in favour of more environmentally-friendly products and services... In the right circumstances, environmental taxes can be highly effective in both cost and environmental terms as the differentiated tax rates on leaded vs. unleaded petrol demonstrated. (European Commission 2001c, p. 15)

A preference by the EU for market-based approaches is also plainly on view within the Treaty of Nice, which states that the 'use of incentives and instruments which are market-oriented' (Treaty of Nice – Article 175) should be employed in order to help promote sustainable development, both at home and abroad.

Thus, EU policies as they relate to the Cotonou Agreement can be assessed from two perspectives. The broader issue concerns the suitability of a weak sustainability approach in seeking to attain the stated Cotonou goals. This vision of sustainable development is one that generates a great deal of controversy and is a case that is yet to be adequately proven. More narrowly, given that the EU has interpreted sustainable development in this manner, it is possible to evaluate the EU's interaction with the ACP on the basis of its own 'rulebook'. This line of inquiry governs the remainder of the book. To what extent, accepting the Polity's own definition as a given, does the Cotonou Agreement appear likely to fulfil its remit?

3
The Rise of the Neoliberal Consensus

Introduction

Before considering the sustainability of EU-ACP relations, it is important to place this relationship against the backdrop of North–South relations in general and the evolution of the 'Washington Consensus' in particular. Relationships such as that linking the EU and ACP do not develop in isolation and must be analysed as part of a wider process. For this reason, it is necessary to investigate the forces governing the interaction between richer and poorer states and consider how these came to be established. It is only once this broader picture has been painted that relations such as those governing the interaction between the EU and the ACP can be adequately understood. Chapter 2 highlights the perceptions colouring the EU's approach to sustainable development; similarly, this chapter outlines the context and evolution of the Polity's vision of a development strategy for the ACP countries within the broader international context. Together, Chapters 2 and 3 allow for a clearer interpretation of the EU's priorities *vis-à-vis* its ACP partners.

In the popular imagination, poverty and the South have come to be almost synonymous. This is particularly true for Africa, where poverty, disease and conflict have taken a heavy toll across the continent. Dictatorial regimes, ever-mounting debts and the rampant spread of HIV/AIDS have created the impression that many African countries are almost beyond help. Yet the picture has not always been one of such pessimism. Following the success of the Marshall Plan[1] in rebuilding Europe after the Second World War, the majority of development theorists were confident that the developing world would be able to play 'catch up' with the wealthier developed economies. Here the various strategies put forward by development theorists are explored and the current state of the debate is assessed.

The Third World, developing countries and the South

Terms such as 'developing nations', the 'Third World' and the 'South' are all used to describe poorer countries and are often used interchangeably. What is problematic with these descriptions is that they imply a homogeneous bloc. The reality is somewhat different. Countries with sizeable populations such as India and China are thus effectively bracketed with the microstates of Mauritius and the Seychelles. Considerable variations also exist where wealth, access to natural resources and degrees of industrial development are concerned. Oil-rich Kuwait and poverty-stricken Mozambique are thus grouped together as developing countries.

The term 'Third World' was coined in 1952 by the economist Alfred Sauvy in an article in the French publication *L'Observateur* and was a reference to the 'Third Estate' of pre-revolutionary France. This description rapidly entered political parlance and additionally necessitated the creation of the terms 'First World' and 'Second World'. First World and Second World were relatively easy to define as members exhibited a number of common economic and political features. In general terms, the First World included the industrialised nations of Western Europe, the USA, Canada, Australia and Japan, while the Second World consisted of the USSR and its Communist satellites in Eastern Europe. All other countries were defined as Third World. However, despite the popularity of this terminology, its use has a number of relatively obvious limitations. From the outset, the countries constituting the Third World were too diverse to ever be considered a bloc. Economically and politically, many of these countries had very little in common. These differences were made even more apparent with the emergence of the Newly Industrialised Countries (NICs) and the 'Asian Tigers' in the 1970s and 1980s. Nevertheless, certain generalisations are possible in that what these countries tend to share is a common history involving colonialism, delayed industrialisation and skewed economic development. At the same time, it is argued in some quarters that as the Third World was defined largely in relation to Cold War divisions, the term has become, in effect, defunct with the collapse of the bipolar world order in 1990 (Bayart 1991, pp. 51–71). In addition, there is a growing perception that the term has become 'loaded', implying stagnation, disorder and oppression, and as such has fallen out of favour with many in the development field, being superseded by 'North-South' and 'developed-developing' which are deemed to be less pejorative.

These alternatives are by no means wholly adequate replacements. The attempt to introduce a distinction between the industrialised North and the undeveloped South has not been altogether satisfactory. Although not strictly geographically defined, this characterisation is problematic for a number of obvious reasons. Mongolia is situated in the North and yet cannot be described as a developed country. Likewise, Australia in the South cannot be bracketed with countries in Africa and Latin America. Similarly, endeavours to define countries in terms of development levels, dividing nations into developed and developing countries, have fallen short. Evidence shows that in many instances countries are not developing at all and are in fact becoming poorer. This in turn has led to the classification of certain nations by the UN as Least Developed Countries (LDCs).[2]

However, while patently not a homogeneous bloc, many of these states do have certain similarities, enough to make distinctions such as 'Third World', 'North' and 'South', 'developed' and 'developing' useful terms. Consequently, these terms do have a degree of use value when attempting to make basic generalisations. Owing to the fact that 'Third World' is now viewed in many quarters as having particularly pejorative connotations it is not employed here. Instead reference will be made to 'developed-developing' countries and 'the North' and 'the South'. These terms will be used interchangeably.

Development paradigms

During the 1960s and 1970s, developing countries averaged an economic growth rate of 3.7 per cent, significantly higher than that of many developed areas, and theorists were optimistic that this growth would continue. However, this figure is misleading, belying a considerable variation within the group. Areas such as Sub-Saharan Africa (SSA) only managed to attain an overall growth rate of 0.2 per cent during this period (Lewellen 1995, p. 73). The two oil crises of the 1970s and the global recession of the 1980s stifled a large percentage of what growth there was, to the extent that, by the 1980s, SSA as a whole actually achieved negative growth. Furthermore, income per capita growth between 1950 and 2005 has, in the case of LDCs, been negligible.

While the statistics highlighted in Table 3.1 cannot offer any indication as to human welfare or standard of living, they nonetheless offer

Table 3.1 Per capita income indicators in US dollars (1969–2005)

	1969	1975	1985	1995	2005
EU(15)	10,775	12,535	15,338	18,519	22,532
USA	18,080	20,069	25,477	29,910	37,651
SSA	553	590	544	492	574

Source: United States Department of Agriculture (2006).

some useful insights. The figures demonstrate, albeit somewhat crudely, how the gap between the rich and the poor has become greater in the past four decades. A number of theories have been put forward to account for this inability of the South to achieve sustained economic development. Growth theory, modernisation theory and neoliberalism all place an emphasis on internal factors being the root cause of the *impasse*. Only dependency theory and its variants argue forcefully that it is external factors which prevent development. However, it has been the neoliberal paradigm that has come to dominate the development debate and its hegemony will be discussed below and contrasted with the other theories.

Growth theory and modernisation theory considered

Given contemporary pessimistic forecasts for African economic development by donor bodies and development agencies, it is perhaps difficult to imagine just how optimistic these same bodies were in the aftermath of decolonisation. The post-war economic boom in the USA, Japan and Europe, as well as the apparent success of Keynesian economic strategies in engendering employment and welfare opportunities, bolstered this sense of optimism. Many African leaders also drew inspiration from the USSR's Five Year Plans in the 1920s and 1930s which seemed to offer a possible blueprint for industrial development. The UN General Assembly declared the 1960s to be a 'development decade' and maintained that the goal for developing countries was to be a minimum of 5 per cent growth in aggregate national income within ten years. This goal was raised to 6 per cent in the 1970s. (Leys 1996, p. 108). It was in this spirit of optimism that early development paradigms such as growth theory and modernisation theory were advanced. Proponents were confident that high levels of sustained economic growth would be possible for these nations and that they would 'catch-up' with the industrialised countries. Initially, analysts believed that low levels of development were a primary condition through which all countries had to pass *en route* to a developed economy (Lewellen 1995, p. 19).

Growth theory evolved from certain Keynesian economic principles. It stressed that state intervention was a necessity if countries in the South were to achieve lasting development. Accordingly, it was maintained that development could not be left to the vagaries of the market, as this was seen to offer little hope where 'kick-starting' the process of industrialisation was concerned. Low levels of savings, poor infrastructure and a lack of investor confidence were believed to make market-driven responses unpractical for these countries (Brohman 1997, pp. 12–13). Walter Rostow's *Stages of Growth* hypothesis represents one of the cornerstones of growth theory. In this work Rostow argued that all countries must pass through five stages of economic growth (see Table 3.2) on the way towards development, each representing a shift from a traditional to a modern society (Lewellen 1995, p. 55).

The crucial aspect of this process was what Rostow (1958, p. 154) described as the 'take-off' phase; the 'push' that set in motion the process of modernisation:

> Take-off is defined as the interval during which the rate of investment increases in such a way that real output per capita rises and this initial increase carries with it radical changes in production techniques and the disposition of income flows which perpetuate the new scale of investment and perpetuate thereby the rising trend in per capita output.

Table 3.2 Rostow's stages of growth

Stage	Characteristics
1. The traditional stage	Production limited by pre-Newtonian science, limited technology and an enforced hierarchical social order.
2. Preconditions for take-off	The emergence of an elite group prepared to invest wealth in the fostering of national and international economies.
3. Take-off	The restructuring of society in such a way so as to pursue increased investment and industrial output.
4. The drive to maturity	A period of consolidation.
5. The age of mass consumption	A shift towards the production of consumer durables and the provision of services.

Source: Rostow (1958, pp. 154–66).

However, Rostow argued that certain preconditions were necessary if take-off was to be successful. The first step in the process required a group within a given society to take responsibility for the introduction of new production techniques. Rostow stressed the importance of society as a whole recognising the value of such changes and responding positively to them. Crucially, positive political, social and institutional responses were necessary in order to free the society of deeply rooted, traditional values. Only once such preconditions were in place could the take-off process become successful. Rostow stressed that the foundation process could take up to a hundred years to be fully enacted. Furthermore, he asserted that take-off itself could continue for up to two to three decades – the system eventually becoming self-perpetuating in the latter stages. If low levels of domestic capital were highlighted as obstacles to take-off, then Rostow advocated government intervention (Brohman 1997, p. 13).

By the late 1950s, several weaknesses in growth theory had become apparent. The theory was problematic in that it assumed that development in the South would mirror patterns observed in the North, despite the enormous differences between them in terms of history and circumstance. Furthermore, the narrow economic focus of growth theory ignored the social and political aspects of development (Brohman 1997, p. 15). Modernisation theory, which evolved out of Rostow's work, can be viewed as an attempt to seek a broader understanding of development. Proponents asserted that capitalist societies were composed of various typical patterns: economic patterns such as consumerism and savings, sociological patterns such as high literacy and urbanisation and psychological patterns such as achievement motivation and rationalism. Countries in the South were urged to adopt similar patterns through the shifting of traditional norms and values so as to better reflect those of industrialised states. Modernisation theorists argued that countries with very low levels of development were in an ideal position to achieve rapid rates of growth. By adopting the best available technology, these countries could circumvent a long, protracted development stage and therefore rapidly 'close-the-gap' between themselves and the developed nations. The greater the differences in wealth, the greater the opportunity for significant advancement (Abramovitz 1993, p. 59). However, for progress of this nature to occur, certain societal norms, such as rationalism, were deemed a necessity. Adherents postulated that catch-up would be most successful in societies that were technologically 'backward' but socially 'advanced'.

The perceived importance of norms and values

The suspicion that norms and values might lie at the centre of the development debate has lingered. Max Weber's work is regularly cited in order to emphasise a link between societal norms, rationalism and the rise of capitalism in the West. For Weber, the growth of capitalism in Europe was coupled with the 'Protestant work ethic'. In his view, Protestantism encouraged self-reliance, hard work and the desire for self-improvement. These values, Weber argued, are what made the growth of capitalist society possible. Modernisation theorists, too, emphasised that, if catch-up was to be attainable, then the society in question should hold values compatible with capitalist development. Herman Kahn, writing in the late 1970s, and a key proponent of what would later become described as 'Reaganomics', continued the debate in a similar vein, stressing the importance of the Confucian ethic to the development of the 'Asian Tigers' (Kahn 1993, p. 169). He went so far as to state that the Confucian ethic might well be better suited to capitalist development than Protestant values, basing his point on the tendency of Confucian societies to place a high value on education, the promotion of sobriety, the desire for accomplishment and dedication to the task at hand. It was also significant to him that these societies emphasise equality and fairness but, importantly, within an institutionalised setting, thus encouraging loyalty and dedication as well as responsibility, which in turn reduces the potential for labour unrest.

In the 1960s, David McClelland (1993, pp. 142–53) made an attempt to demonstrate a link between economic growth and cultural values. He studied a broad cross-section of children's stories from across the world on the basis that these tend to change little over time and therefore provide a good indication of the values with which a society seeks to imbue its offspring. His findings suggested that societies which have traditionally stressed the importance of achievement also demonstrate the highest levels of economic development; he cited the Protestant work ethic as being indicative of this trend. However, McClelland was at pains to emphasise that values are not constant and that they evolve over time. Theoretically, then, all societies are capable of instilling values conducive to economic growth within their cultural mindset.

This theme was carried through to the 1980s, with Lawrence Harrison (1993) discussing the extent to which 'underdevelopment is a state of mind'. He argued that worldview determines society's values and, as such, is the key to understanding why some societies have not achieved economic development. Furthermore, he stressed that it is important that a country's worldview be forward-looking and not fixated on past or even present glories. Societies caught in the past, he maintained, cannot plan for the future. Attitudes towards rationality are also crucial. He stated that if a society believes that the natural world can be explained, then it is more likely to accept science as a tool for development. Superstitious societies relying on magic and portents to define their existence will tend to be wary of science due to its potential threat to their belief system. Harrison (1993, p. 178) argued that it is no accident that the majority of Nobel Prize winners have emanated from societies which base their worldview on rationalism. In his view, increased levels of development could be achieved by a collective change in mindset.

Dependency theory

The optimism that fuelled development theorists in the 1960s and early 1970s was replaced with pessimism as the severity of the 'African Crisis' became increasingly apparent in the early 1980s. This growing sense of pessimism was most clearly articulated by the dependency school. Dependency theory developed as a critique of modernisation theory and its related ideas, as outlined above, and until the 1980s exerted a significant degree of influence over the North–South debate. However, the collapse of communism, internal contradictions and structural weaknesses meant that it is no longer the force it once was. Nevertheless, the theory provides a strong critique of modernisation theory and for this reason alone is worth addressing (it is also, arguably, an implicit theory of globalisation. This point is discussed in more detail below). In contrast to previous models, dependency theorists argued that external factors significantly outweigh internal ones. Of singular importance is the historical position of any given country in relation to that of the industrialised economies.

Dependency theory grew out of the Latin American experience, largely inspired by the influential work of Fernando Henrique Cardoso and Enzo Faletto (1979), in the mid-1960s. However, it was André Gunder Frank (1967), strongly influenced by Paul Baran (1957), who was largely responsible for the dissemination of dependency theory beyond the confines of Latin America. Frank maintained that industrialised countries achieved development only by deliberately 'underdeveloping' the areas they colonised. The term 'underdeveloped' was coined by Frank in 1967 (pp. 3–12) as a means of explaining the relationship between the coloniser and colonised. His work from this period argues that colonies were not left undeveloped but, rather, were underdeveloped in so far as their economies were based on what was most beneficial to the colonial powers. Viewed in this manner, development and underdevelopment are thus two sides of the same coin. For the one to occur, the other is a necessity. Concepts such as 'centre' and 'periphery' were also introduced in order to better describe the relationship between developed and underdeveloped regions. Dependency theory therefore argues that the economies of the periphery, lacking an internal dynamism, are shaped by external factors.

Frank (1996) argued that the historical experience of underdeveloped countries differed markedly from that of developed countries, therefore making generalisations as to 'stages-of-development' and 'growth patterns' an impossibility. Frank stressed that Rostow's 'traditional stage' existed only in theory, as no part of the globe has been untouched by capitalism. If this is the case, then the foundations for Rostow's theory are wholly undermined. In addition, since orthodox theories fail to take into account the historical processes with which the global economy has been shaped, they offer little insight into the causes of underdevelopment. Inequalities can only be explained through a detailed understanding of the history of capitalist expansion. Any development occurring in these countries has effectively been satellite development, neither self-generating nor self-sustaining. For Frank (1996, p. 108), inequalities were dependent not on structural inadequacies, cultural shortcomings or lack of funding but rather on the historical processes which have existed since the birth of the capitalist economic system. He stressed that when underdeveloped states have tended to make progress, this has occurred when links to the centre have been weak. The First World War, the Great Depression and the Second World War were periods of significant growth in Latin America, especially in Brazil, Chile and Argentina. Frank asserted that these periods, during which the centre had little choice but to loosen ties with the periphery, offered the best opportunity for a degree of autonomous development. If any lesson is to be learned from their experience then it is that independent growth (autarky) is essential (Rosen and Jones 1979, p. 247).

Frank's work was, however, largely confined to the Latin American experience and it was theorists such as Walter Rodney (1982) and Samir Amin (1972) who provided the dependency debate with a more ACP-relevant emphasis, focusing on the extent to which colonialism underdeveloped the African continent. The most important aspect of the colonial relationship was, for Rodney (1982, p. 225), that which centred on issues of power:

> The colonisation of Africa lasted for just over seventy years in most parts of the continent. This is an extremely short period within the context of universal historical development. Yet, it was precisely in those years that in other parts of the world the rate of change was greater than ever before ... The decisiveness of the short period of colonialism and its negative consequences for Africa spring mainly from the fact that Africa lost power ... When one society finds itself forced to relinquish power entirely to another society, that ... is a form of underdevelopment.

Colonialism destroyed the political infrastructure in Africa and left in its place puppet rulers and appointed chiefs. Rodney argued that being stripped of the ability to act in a meaningful sense resulted in the colonised being effectively 'removed' from history, becoming passive objects rather than agents active in the shaping of the continent. Furthermore, once in control, the European powers dictated the course of economic development. Colonial infrastructure was developed to facilitate extraction, and all trade links were focused on the 'mother' country. Africans were required for their labour alone, rather than their expertise. The result, for Rodney (1982, pp. 230–31), was the stifling of indigenous technological development and innovation. Cash-crop farming, enforced migrant labour and the introduction of a monetary economy all served to erode traditional cultural ties and norms. This was particularly true for southern Africa, where extensive mineral extraction was dependent on migrant labour. Decolonisation did not end underdevelopment, as unequal power relations resulted in a system of 'neo-colonialism' whereby the metropole continued to dictate to the periphery.

Dependency theory has been criticised as self-defeating, in that it offers little by way of a solution – a point readily acknowledged by Frank (1991, p. 32) when writing in the early 1990s:

> Real dependence exists, of course, and more than ever despite denials to the contrary. However, dependency 'theory' and policy never answered the question of how to eliminate real dependence and how to pursue the chimera of non- or in-dependent growth.

Dependency theory also stands accused of being overly reductionist and thus incapable of producing an accurate picture of the causes of underdevelopment. Frank focused mainly on macroeconomics, and as a result, largely ignored the political process; he also failed to allow for any significant autonomy in the creation of economic policy. By ignoring the state as the unit of analysis, he dismissed any possibility of an independent role for individual governments, political parties and state bureaucracies, dismissing them all as mere phenomena of the overall system. In addition, from an empirical point of view, dependency theory is weak in that there is little reliable data with which to reinforce its assertions. However, it was the emergence of the NICs, and especially the 'Asian Tigers', that significantly undermined the dependency position. South Korea, for example, had endured four decades of oppressive Japanese rule, the debilitating effects of the Korean War and decades of corrupt government before embarking on a period of sustained economic growth. Between 1965 and 1990 its economy achieved an average growth rate of 7.1 per cent, to becoming, by the late 1980s, the world's eleventh largest economy (Randall and Theobald 1998, p. 138). Furthermore, this growth was achieved by indigenously owned *chaebols*, conglomerates centred on a single holding company, such as Samsung and Hyundai, all of which suggest that genuine development in the face of historical dependency remains a possibility.[3]

Dependency theory has also been criticised for being historically weak. The historical basis for the dependency argument is not sufficiently universal to provide the theory with substance, and reducing of the influence of the North to a continual process of surplus extraction means that nuance is sacrificed to generalisation. Nevertheless, many of the issues raised by dependency theory, including high levels of indebtedness, an emphasis on the export of primary commodities, and the opening up of domestic markets, remain pertinent to any discussion of development.[4] Many of these issues are now commonly addressed within a framework of globalisation theory (discussed below), which offers a more flexible approach to analysing such matters than that offered by dependency theory. However, the fact remains that, as stated by Walter Rodney, 'the vast majority of Africans went into colonialism with a hoe and came out with a hoe' (cited in Leys 1996, p. 111). The clear lack of technology transfers or capital investment gave dependency theory a strong 'commonsense' appeal, especially when viewing North–South relations in terms of exploitation. Dependency theory was also responsible for the 'demystification' of the African development experience, stressing the historical and social nature of that experience. It also served to highlight the explicit and implicit 'strings' attached to aid, the influence of foreign capital over political leaders, and the fact that African development was profoundly influenced by the requirements of the capitalist system. It is for this reason that much of the language of dependency theory has endured and is still regularly employed by African statesmen (for example, see Mbeki 2003).

Neoliberalism

Both modernisation and dependency theory have now been effectively displaced by the resurgence of the neoliberal economic paradigm and the establishment of the so-called 'Washington Consensus'.[5] Prior to the global economic recession of the early 1980s and the end of the Cold War, countries in the South had a certain number of economic strategies available to them, as well as a variety of aid donors. The politics of the Cold War ensured that finance was readily available, and developing countries took advantage of such funds in order to bolster their economic policies. However the recession caused by the two oil shocks of the 1970s resulted in a cessation of such funds. As the global economic slowdown started to bite, many countries found that they could no longer service the extravagant loans that they had secured during better times. Most had no option but to turn to institutions such as the International Monetary Fund (IMF) and World Bank in order to remain solvent.

During the 1960s and 1970s, borrowing on a grand scale by developing countries was encouraged by the international community in order to offset the former's low levels of savings. Borrowing was seen as a strategy necessary to ensure the continuation of growth. During this period, primary commodity prices, from which most countries in the South derived the bulk of their income, were high. Debt servicing was thus not an important issue. However, the first oil crisis in 1973, which saw oil prices quadruple, meant that poorer countries were forced to borrow further in order to sustain their economies. Between 1975 and 1980, external debt in the South rose by 353 per cent and yet, as commodity prices remained high, servicing was not overly problematic (Newark 1995, p. 231). However, the second oil crisis in 1979 resulted in the debt levels of developing countries spiralling out of control. Global commodity prices collapsed while interest rates rose, thereby creating ever-mounting levels of debt. In 1982, Mexico heralded the onset of the 'debt crisis' when it declared that it could no longer service its obligations.

The global recession also impacted heavily on the developed economies, all of which experienced economic downturns. The demand for a fresh economic impetus led to the election of conservative governments in the USA, Britain, Canada and West Germany, all of which favoured neoliberal policies. Keynesian strategies were gradually abandoned in favour of fiscal austerity, reduced government participation in the economy and general economic liberalisation. These policies had a knock-on effect on donor programmes, which rapidly assumed a neoliberal stance. Whereas dependency theory laid the blame for underdevelopment on global forces, the neoliberal paradigm implicated state intervention and institutional weaknesses as the key impediments to growth in the South.

In 1981, the World Bank, under the leadership of Robert McNamara, published the 'Berg Report' which called for a 'rolling-back' of the state in African countries, privatisation and greater liberalisation. The Report (World Bank 1981) signalled the advent of what would come to be known as 'structural adjustment'. The IMF followed the World Bank's lead in this area, introducing Structural Adjustment Programmes (SAPs) in the early 1980s, with Mexico serving as an initial, high-profile, test case. SAPs were based on the principle of economic restructuring and differed little from countrytocountry. The core goals inherent in SAPs included cutting government expenditure, curbing inflation, liberalising trade, privatisation, devaluing artificially high currencies and removing price controls. These reforms became compulsory for developing countries hoping to succeed in securing aid from the IMF. As early as the mid-1980s, two-thirds of African countries were being financed by the IMF, while three-quarters of Latin America was being similarly funded.

In many respects, and where Africa in particular is concerned, government reform has been long overdue. The inefficiencies of the post-colonial African state apparatus were apparent to observers from the beginning. While few analysts chose to take any notice, the economist René Dumont noted grave institutional weaknesses in the former French colonies almost immediately after their independence. In *False Start in Africa* (Dumont 1988), he railed against the bloated civil services in the Francophone states, complaining bitterly of exorbitant salaries, corruption, patronage and the power of domestic elites. He asserted that the principal 'industry' of these countries was bureaucracy, a condition which would, he warned, be their ruin. His was a minority view. Most Western governments were extremely well disposed towards these new states and pessimistic forecasts were generally dismissed as racist. However, within a decade of independence, marked weaknesses in African states became apparent, both in terms of legitimacy and their ability to govern effectively. Patrick Chabal (1997; Chabal and Daloz 1999) maintains that the patrimonial nature of the African state is the root cause of its weakness. In precolonial times, African societies were, byandlarge, governed within a framework of client–patron relationships. The system was often utilised by the colonial rulers in their drive to implement forms of indirect rule. With independence, most states adopted Western-style nation-state political institutions. However, these were often at odds with the prevailing political culture, which still relied heavily on patrimonial relationships. Rulers thus had to adapt political institutions in order for them to be viable. Chabal (1997, p. 8; 1999, p. xix) describes this method of government as a post-colonial form of patrimonialism (neo-patrimonialism). In order to establish ties of patronage, ruling elites created one-party states so as to control resource flows. This placed the state at odds with sectors that were excluded from the patrimonial relationship, effectively forcing them to challenge the legitimacy of the former.

The nature of the African state has prevented many of these countries from adapting to the 'new economic order'. Ruling elites have tended to finance their regimes through high levels of government spending, including resorting to subsidies in order to artificially depress the price of staple commodities. The majority have also distorted the true cost of healthcare and education. In many instances the state has also controlled key sectors of the economy by virtue of the nationalisation of key industries. International donor bodies now routinely demand a complete reversal of almost all of these policies. As a result many regimes soon found themselves under immense pressure. Furthermore, as a demonstration of their commitment to good governance, formerly one-party states were compelled to hold democratic elections. These measures presented a number of problems for the ruling elites. By removing subsidies for staples, in some instances as high as 85 per cent of the true cost, governments risked the wrath of an urban population galvanised by dramatic price increases (Bates 1993, p. 335). By reducing the size of the civil service, governments jeopardised their ability to dispense patronage, thereby risking the alienation of previous supporters. In the wake of these unpopular decisions, the demand for democratic elections was one that caused considerable unease.

The social costs of economic reform have also been high. The introduction of user-fees for health and education has priced these commodities outside the reach of many people. In Ghana, primary schools recorded a 40 per cent drop in attendance after the introduction of school fees. Hospital visits fell by a third. In Uganda too, the privatisation of nationalised industries resulted in the loss of 350,000 jobs (Cromwell 2001, p. 50). In addition, with the removal of government subsidies and currency devaluations, the cost of basic goods in these countries soared. Immediately following a currency devaluation in Sierra Leone in 1986, the price of kerosene rose by more than 100 per cent, the cost of soap by 300 per cent, and firewood by nearly 175 per cent (Riddell 1995, p. 216).

Most African states are now effectively tied to the IMF through structural adjustment or other forms of conditionality. While structural adjustment policies have been in place for a number of years, it is difficult to find evidence for any dramatic improvements in people's living standards. Ghana and Uganda are often portrayed as IMF-World Bank success stories. The World Bank claims that the economies of these two countries have grown rapidly since the reformation of their policies, this despite the fact that they have not received as much aid as some other African nations. The reason that their programmes have been successful, it is argued, is that the leaders of Ghana and Uganda have embraced reform and have determinedly followed it through. This success, however, is relative. First, both countries turned to the IMF and World Bank only because they were in the midst of severe economic crises. In 1983, Ghana was experiencing inflation of over 100 per cent and Nigeria was threatening to expel all Ghanaians (1.2 million) from within its borders, thereby exacerbating the growing sense of instability. Similarly, by 1986, Ugandan inflation was running at over 300 per cent and the country's financial services sector was on the verge of collapse (World Bank 2001, p. 7). Both Ghana and Uganda were on the brink of bankruptcy. Despite their 'success' in the eyes of the global economic establishment, both are heavily indebted and income per capita remains low. Real reductions in poverty rates remain elusive.

Furthermore, the neoliberal development model is based on a number of dubious premises. The IMF points to the Asian Tigers as examples of the successes that can be achieved through export-led growth. Yet, closer examination of the economic development of these countries suggests that their growth was not determined solely by market principles or liberalisation. In both South Korea and Taiwan, the state was actively involved in the economy. Wages were artificially depressed, the labour movement was kept weak and the government offered vast subsidies and tax breaks as a means of stimulating key economic sectors. In addition, punitive tariffs kept foreign competition from undermining local enterprises. However, the state in both cases pushed hard to develop export markets for their manufactured goods. Consequently, they established a two-prong policy of protecting domestic markets whilst advocating free trade for their exports, with the result that in effect, as Nigel Harris puts it, 'the invisible hand became an iron fist' (cited in Lewellen 1995, p. 109). Both South Korea and Taiwan developed as a result of carefully constructed policies of economic nationalism. The manner of their development is frequently glossed over by the IMF/World Bank analysts who tend to focus only on their export-led economic growth.

Furthermore, developed countries tend to overlook the fact that they too still employ a vast array of subsidies and tariffs. While state intervention in the economies of the developing countries has in many instances been disastrous, this factor alone cannot account for the malaise in the South. Between 1950 and 1975, developing countries experienced high levels of growth despite active government intervention in the economy. South Africa, for example, diversified its economy behind a wall of tariffs and through generous government subsidies. By making use of import-substitution strategies, the South African Government was able to achieve a sustained GDP growth rate of 6 per cent in the 1960s, thereby forming the platform for future economic development (Lewis 2002, p. 2). External factors such as the oil crises and collapsed commodity prices must therefore also be taken into account. Consequently, it is arguable that it is too simplistic to reduce poverty in developing countries to a number of state-sponsored market failures.

There are signs that the paradigm is beginning to soften, and there is now a basic consensus that, particularly in the case of LDCs, some degree of debt relief, or even debt cancellation, is necessary. The UNICEF publication of *Adjustment with a Human Face* in 1987 highlighted a number of the welfare inequities associated with SAPs (Cornia *et al.* 1987). The World Bank accepted a number of the criticisms levelled at its policies in developing countries and duly set up 'social dimensions of adjustment' (SDA) programmes to compliment its SAPs, which in turn morphed into talk of poverty alleviation strategies. Accordingly, in 1996, the IMF and World Bank jointly launched the Heavily Indebted Poor Countries (HIPC) initiative. The initiative brought together bilateral, multilateral and commercial creditors in order to facilitate debt relief for some of the poorest countries in the World (O'Brien and Williams 2007, p. 325–26). The scheme came with conditionality – debt relief in exchange for meeting proscribed targets. After an unsteady start, the programme has demonstrated that the HIPC has channelled substantive additional development resources to qualifying countries (World Bank 2006b, p. 34). The HIPC initiative was followed up in 2006 by the Multilateral Debt Relief Initiative (MDRI). The MDRI, unlike the HIPC, allows for 100 per cent debt relief for countries that have borrowed from the IMF, World Bank or African Development Bank (O'Brien and Williams, 2007, pp. 326–27). The result is what some commentators (see Stiglitz 1998) have described as a 'post-Washington Consensus' which includes endeavouring to convey the importance of social justice, quality of life, and environmental security, as well as economic growth. Nevertheless, despite these changes, the basic principles of the neoliberal paradigm remain entrenched and it is difficult to conceive of any realistic challenge emerging from other quarters in the near future.

Globalisation

Closely tied to the neoliberal paradigm is the phenomenon of globalisation. Since the abandonment of Keynesian economic policies at the beginning of the 1980s, there has been a shift towards increased interdependence and internationalisation, particularly where the economic sphere is concerned. Free-flowing capital, labour and technology are all indicative of this new economic order. There is a growing perception that the era of the nation state as an autonomous entity is over; that in the face of a globalised economy the role of the state is that of a local authority, supplying infrastructure and stability.[6] Certainly, economies are now geared towards the global market rather than towards any domestic demands, and it is for this reason that international donor bodies require countries in the South to liberalise their economies, that they might become active participants in a global economy based on neoliberal principles. However, the rhetoric surrounding the issue of globalisation is problematic. Globalisation is presented as both a new paradigm and a new economic system, differing markedly from its predecessors and thus not subject to the same pressures and constraints. Closer inspection suggests that this is not necessarily the case.

Paul Hirst and Grahame Thompson (1999) argued that globalisation is not a new phenomenon. Their evidence suggests that the current internationalisation of trade is not unique and that, in fact, between 1870 and 1914, the global economy was a great deal more 'open' than it is currently. The markets of Britain, Japan and the Netherlands all had lower tariffs before 1913 than they did in 1995. Of all the developed countries, it is only the USA which is now significantly more open than it was before the First World War. The evidence provided by Simon Kuznets in the 1950s reinforces this argument. He demonstrated that, by 1913, foreign trade proportion stood at 33 per cent, a figure that is yet to be surpassed (cited in Hoogevelt 1997, p. 70). In terms of active participation in the global economy, currently marginalised countries were more involved during the colonial period than they are at present. Through colonialism, the South received up to half of all direct foreign investment. By 1989, this had dropped to just 16.9 per cent (Hoogevelt 1997, p. 77). States in the South were important contributors to the global economy during colonialism – many no longer are. Furthermore, Hirst and Thompson (1999) argued that internationalisation is increasing, the increase is by no means evenly spread. A triad comprising North America, Europe and Japan accounts for the bulk of foreign investment and all the other hallmarks of globalisation. Whilst countries such as China, India, Brazil and other G-20 developing countries are now increasingly powerful and vocal actors in the global economic arena, the vast majority of the G-90 countries contribute little to the system and likewise receive little from it.[7]

Hirst and Thompson criticised what they perceived as an over-emphasis, on the part of many 'strong' globalisation theorists, concerning the supposedly increased irrelevancy of national borders to the structure of economic flows. Only Western Europe, they argued, is anywhere near approaching an economically integrated zone and even in this instance the level of integration in terms of GDP traded between states is lower than might be expected.[8] While the expansion in international growth is studied in great depth, few theorists have noticed the parallel growth in domestic trade. This suggests that national boundaries remain an important vehicle for the stimulation of economic growth. If this is the case, then governments are capable of influencing economic policy to a far greater extent than has been suggested by those stressing that national economies are dissolving in the face of economic globalisation.

Jan Aart Scholte argues that what is 'new' about the globalisation process is the evolution of 'supraterritoriality'. He infers that 'social space is no longer wholly mapped in terms of territorial places, territorial distances and territorial borders' and this, for Scholte (2000, p. 16), is a new phenomenon. As has been mentioned, liberalisation and internationalisation are not without precedent. However, supraterritoriality is a recent development. Until the 1960s, social geography had always been tied to specific territories. People traditionally defined themselves in relation to their place within a particular location, often a nation state, with most social interaction taking place within the confines of this territory. However, late-twentieth-century technological advances began to alter people's perceptions of space. Electronic finance, international telephone calls, satellite television and the internet are not tied to national boundaries and consequently engender a view of the world as a single entity. Granted, certain religions entertained this view centuries ago, with Christianity and Islam certainly viewing themselves as 'world' religions. However, it is only since the Second World War that the idea of the world as a single unit has become conceivable. Much of the developing world is, however, at present excluded from this form of globalisation, its citizens lacking the technology necessary to make this supraterritorial world a reality.

Sociologist Gary Gereffi (1997) maintains that the picture for developing countries is not as bleak as might be inferred from the above. He proposes that, since the 1960s, a more integrated global division of labour has occurred in both production and trade and that, consequently, new patterns of specialisation have occurred. In order to gain from the process of globalisation, the developing countries need to attract capital in search of cheap labour. During the 1960s and 1970s, the Asian Tigers initially lured investors to their shores in part due to their ability to provide cheap, yet relatively skilled labour. A rapid rise in real wages and dramatically strengthened local currencies have since largely blunted this advantage. Countries able to offer low labour costs coupled with efficiency and productivity will always be attractive to investors. Gereffi argues that the classic 'core-periphery' model (dependency theory), in which the periphery provides primary goods while the core exports manufactured products, is outdated. Instead, as developed countries have shifted the focus of their economies to the service sector, so manufacturing has begun to account for much of the South's income. Through technological development and, by learning from the South East Asian success stories, Gereffi argues that developing countries can successfully couple themselves to the global economy.

The perceived benefit of globalisation for developing countries is that it creates wealth and, accordingly, reduces poverty. This is a view which is strongly backed by World Bank analysts. David Dollar and Paul Collier (2001, p. 1), for instance, maintain that 'globalisation generally reduces poverty because more integrated economies tend to grow faster and this growth is usually widely diffused'. They divide developing countries into 'globalisers' and 'non-globalisers'. Globalisers, countries which have made efforts to reform their economies and participate in the global economy, have experienced growth rates of approximately 5 per cent over the past decade. Countries such as China, India, Mexico and the Philippines are included in this category. These are highly populous countries representing a significant percentage of the global population, with three billion people now currently experiencing sustained economic growth as a result of their reform agendas. Dollar and Collier argue that, for the countries that have not embraced globalisation, the prospects appear bleak, with non-globalisers incorporating over two billion people who face the prospect of being permanently confined to the margins of the global economy. The solution, the World Bank asserts, is greater liberalisation, not only for poor countries but also for developed nations. The World Bank also emphasises the need for all countries to further reduce remaining trade barriers. Developing countries should strive not only to liberalise their economies but also to create a positive environment for investment through good governance. This, coupled with better educational services, is seen as the best means of encouraging capital flow (Dollar and Collier 2001, pp. 10–12). However, Dollar and Collier fail to touch on the experiences of the Asian Tigers when discussing the success of the globalisers, and neither do they address the impact of the state in cementing these countries' economic fortunes.

Despite such problems, globalisation and neoliberal economic policies appear to be viewed by both proponents and opponents as somewhat inevitable. There is a sense that globalisation is an unstoppable, dynamic and very deterministic force against which resistance is futile. Other paradigms are thus often ignored as a result of the belief that globalisation is a *fait accompli*. At a WTO ministerial meeting in 1998, the then US President, Bill Clinton, stated that as far as he was concerned 'globalisation is not a policy choice – it is a fact' (cited in Cromwell 2001, p. 9). However, Hirst and Thompson (1999, p. 7) argued that, as gobalisation is essentially not a new phenomenon, regulation of the international economy was still possible. They stressed that globalisation was merely 'a conjectural change toward greater international trade and investment within an existing set of economic relations'. Furthermore, not only is globalisation not inexorable, but historical precedents suggest that the process could be stalled or even reversed. As mentioned, First World War brought an end to the 'globalised' era which preceded it, ushering in an era of heightened protectionism which was only partially dismantled following Second World War. Hirst and Thompson claimed that a similarly dramatic reversal in globalising trends remains a distinct possibly, particularly if benefits are not evenly spread. The same controls that regulated markets in the past could still be put to use in this 'new' global era. If Hirst and Thompson were correct, then more can be done to boost the economies of developing countries than merely liberalising their markets. This contention is a challenge not only to neoliberal orthodoxy but also to related perceptions of sustainable development. If true, and if regulation and public sector involvement are in fact successful in facilitating poverty alleviation and environmental protection, then these strategies can be seriously considered as components of a wider development paradigm when merited, rather than simply dismissed as failed approaches from a by-gone era. As will be demonstrated in subsequent chapters, when markets are weak or non-existent, applying neoliberal economic precepts can in fact be counter-productive.

Conclusion

Until the early 1970s, it was widely believed that poverty in developing countries could and would be eradicated. Economic growth in the South reached impressive levels between 1950 and 1975 and there was no reason to suspect that this growth would not continue indefinitely. Both growth and modernisation theorists were optimistic that a 'catch-up' was possible. However, by the early 1980s these hopes had been dashed and the credibility of related development paradigms was in doubt as the neoliberal model began to assert itself. The dominance of the latter in determining IMF, World Bank and other donor policies has ensured that the majority of developing countries have been forced to accept its tenets. The success rate of such policies, particularly in Africa, has been negligible, often resulting in insignificant growth rates coupled with increased unemployment and higher levels of illiteracy. However, the neoliberal paradigm is well established and while there have been measures taken by the emerging 'post-Washington consensus' to soften its edges, such as promises of debt forgiveness and poverty alleviation strategies, the basic framework remains intact. Furthermore, the rapid spread of globalisation has worked to cement the dominance of neoliberal economics. In many respects, despite the arguments of Hirst and Thompson, neoliberal, economically driven globalisation is viewed as an irresistible force, and one which developing countries can ill-afford not to embrace. As no viable alternative strategies acceptable to the developed countries exist, it is inevitable that the South will have to attempt to adapt to the 'new world order'. Many countries in the South are clearly unhappy with the prospect of a 'one-size-fits-all' solution to their economic concerns. This dissatisfaction was plainly demonstrated in multilateral negotiations such as the WTO summits held in Seattle (1999) and Cancun (2003) and the continued failure to bring about a closure of the Doha Round.

The relentlessness of the neoliberal agenda coupled with the 'inexorable' progress of globalisation is mirrored in the EU-ACP relationship. Colonial links have ensured a lasting relationship between the two blocs that has endured from the earliest days of the EU. However, the end of the Cold War and the collapse of communism have heralded a new era in this relationship. The patterns indicative of the post-1990 world order are clearly present in all EU-ACP interaction. Those who fear neo-imperialism are particularly concerned about the apparent reversion by the dominant states to old-style, nineteenth-century geopolitics (Klare 2003, pp. 4–5). It is therefore important to assess the ability of the ACP states to act in the face of EU dominance. A pivotal aspect of the Cotonou Agreement is its shift towards WTO-compatibility and the establishment of a new era of cooperation based on largely neoliberal precepts. The decision by the EU to tailor the Cotonou Agreement along WTO lines can thus be viewed as an extension of the globalisation process. The chapters that follow focus on the extent to which this largely neoliberal agenda is capable of delivering sustainable development and poverty alleviation amongst the ACP states.

4
The 'Everything But Arms' Initiative

Introduction

Building on the special provisions for LDCs provided by the Cotonou Agreement, in February 2001 the EU approved the implementation of the 'Everything But Arms' (EBA) initiative. The initiative proposed 'granting duty-free access to imports of all products from LDCs, except arms and munitions, without any quantitative restrictions (with the exception of bananas, sugar and rice for a limited period)' (European Commission 2006a). Furthermore, the terms of the agreement were non-reciprocal, allowing LDCs to continue to collect levy tariffs on EU exports. The announcement, coming as it did in the build-up to the WTO Ministerial Conference in Doha, generated an enormous amount of positive publicity for the EU, publicity the EU did its best to fan. The then European Trade Commissioner Pascal Lamy stated in a press conference that

> It's a worldwide first. At the end of the day, we will have 100% access, with no exclusions, except of course for arms. We have delivered on our fine words. This sends a signal to the rest of the world that we are serious about getting the most disadvantaged to share in the fruits of trade liberalization. (European Commission 2001b)

The EBA initiative is important for a number of reasons. It has allowed the EU to claim the moral high ground in the developed world by virtue of its 'pro-poor' approach, allowing the Polity the kudos of 'laying down the gauntlet' to countries such as the USA and Japan. EBA is important too, for LDCs as it affords these countries 'special case' status and offers rich financial incentives to those best placed to exploit the EU's largess.

However, the initiative is also significant where other factors are concerned. In implementing the policy, the EU has effectively split the ACP bloc into LDCs and non-LDCs. Furthermore, it incorporates non-ACP LDCs, thereby further undermining the unity of the ACP grouping. It also serves to acknowledge that, while some countries do indeed represent 'special cases', these anomalies can be accommodated within a largely neoliberal global economic framework. Given that the EBA has been in effect since 2001, it is possible to assess the EU's commitment to poverty alleviation and sustainable development in the 50 poorest countries in world. It also provides an inkling as to how the EU will proceed with development during the Cotonou period.

A brief overview of the initiative

The origins of EBA can be found in the aftermath of the WTO Ministerial Conference in Singapore in 1996 where developed countries had pledged to do more to open their markets to LDC produce. EBA diverted from the EU's Generalised System of Preferences (GSP) in a number of ways. It afforded all LDCs duty and quota free access to EU markets for practically all products, whereas GSP regulations tended to be more country/product specific. Furthermore, unlike GSP provisions, EBA preferences were not to be subject to time limits. The initiative affects 919 tariff lines, excluding just 25, all related to arms. The opening up of three 'sensitive' sectors, namely bananas, rice and sugar, were to be staggered so as to mitigate any undue 'shocks' to the existing regime. As stated, this lowering of tariffs was to be non-reciprocal; LDCs would be able to continue to collect duties on EU exports.

The category of LDCs was created by the United Nations in 1971 and incorporated 25 countries. The criteria were based on having a GDP per capita of less than $100 (in 1968 values – now $750) and on economic vulnerability together with low levels of 'human resources' (for more details regarding classification refer to Appendix II). The number of LDCs has now risen to 50, of which 34 are African countries. All but nine of these states are signatories to the Cotonou Agreement. LDCs account for approximately 10 per cent of the world's population, yet account for only 0.5 per cent of global exports, a drop of 0.3 per cent since 1980 (*The ACP-EU Courier* 2001, pp. 14–16).

Despite its non-reciprocal elements, the initiative conforms to WTO rules. In 1947, when the GATT was negotiated, there was no provision made for 'special cases' such as the LDCs, especially since such categories did not exist at that point. It was only after the completion of the

Tokyo Round in 1979 that the Decision on Differential and More Favourable Treatment, Reciprocity and Fuller Participation of Developing Countries (referred to as the 'Enabling Clause') was adopted. This decision allowed for the 'special treatment' of LDCs by developed countries. The Enabling Clause was carried over from the GATT to the WTO in 1995. Thus, in terms of legality, the EBA initiative adheres to paragraph 2 (d) of the Enabling Clause as well as the provisions of the Cotonou Agreement.

On the face of it EBA represents a significant development as far as the LDCs are concerned. The initiative meant that 919 new tariff-free lines were added to previous GSP agreements. This is potentially significant because the EU is already the largest outlet for LDC exports, absorbing up to 50 per cent of their output in some cases (McQueen 2002, pp. 101–10). Where agricultural exports are concerned, this figure rises to 70 per cent (Cernat *et al.* 2003). Consequently, the initiative was proclaimed by the EU to be a positive step towards the integration of LDCs into the global economy. EU officials emphasised their hope that the USA, Japan and other developed countries would follow the EU's lead and open up their domestic markets to LDC exports.

However, EBA has not always been the public relations triumph for which the EU had hoped. The Commission had originally proposed all LDC products have immediate duty-free access and that even sensitive commodities such as sugar, rice and bananas only be afforded a three-year transitional period before also becoming subject to EBA regulations. An outcry from various sectors within the EU agricultural lobby, in particular the sugar sector, resulted in a degree of infighting within the Commission, with DG Agriculture demanding greater protection for sensitive sectors, whilst DG Trade pushed for the fastest possible transition to full tariff-free access for the LDCs. The outcome was a distinct dilution of the original proposal. Consequently, it was decided that there was to be open access for all products with the exception of sugar, rice and bananas, which were to have their tariffs removed gradually over a period of time. The full liberalisation of the banana trade came into force in 2006, whilst sugar and rice are due to be tariff-free by 2009.[1] This exclusion of agricultural produce has led to calls by some African countries to re-christen the agreement 'everything but farms' (Mutume 2001, pp. 8–9). Additionally, certain ACP countries, such as the sugar-producing Caribbean states, fearing that they will be disadvantaged by the agreement, have described EBA as 'taking from the poor and giving to the poorest' (Action for Southern Africa 2001a). An ACP-EU Joint Parliamentary Assembly, held in March 2001 in Gabon, stated that the body was dissatisfied with the manner in which EBA had been devised. The Assembly felt that there was little consultation between the EU and the other stakeholders and that, as a result, the policy could well have negative consequences for those affected. To remedy the situation it was suggested that the Commission compile impact studies, for both LDCs and non-LDCs, as to the consequences of EBA (EU-ACP Joint Parliamentary Assembly 2001). The resolution emphasised a determination not to allow the initiative to undermine existing sugar agreements and thereby impact negatively on other ACP countries. The fear that EBA could lead to a lowering of prices for agricultural exports from developing countries was also expressed.

The concerns of the ACP countries not party to EBA are warranted and need to be addressed. When EBA was first announced, sugar interests in the EU produced statistics suggesting that the EU would be flooded with cheap LDC sugar. One report, written for the Commission (2000d, pp. 15–16), estimated that LDC exports of sugar to the EU could potentially increase from approximately 100,000 tonnes to 2.7 million tonnes per year. Studies by the European sugar sector maintain that, with careful planning and full utilisation of EBA incentives, LDC sugar production could reach 4.3 million tonnes per year (Association Des Organisations Professionnelles Du Commerce Des Sucre Our Les Pays De l'Union Europeenne 2001, p. 13). However, it is arguable that these figures are grossly inflated. Together LDCs produce just 2.1 million tonnes of sugar annually whilst consuming 3.6 million tonnes (Réseau Europeén de Sécurité Alimentaire 1999, p. 3). While a small number of countries such as Sudan, Zambia and Malawi are net exporters of sugar, the vast majority rely on imports to meet their needs. Even if the net production capacity of sugar-producing LDCs was to increase dramatically, any impact on European markets would be negligible. Total sugar exports from LDCs in 1997 only amounted to 190,600 tonnes, of which half was exported to the EU (Stevens and Kennan 2001, p. 14). Since 2001–02, LDC sugar exports to the EU have risen from 74,185 tonnes to an estimated 190,113 tonnes in 2006–07 (LDC Sugar Group, 3 August 2007). This represents a fraction of the estimated 16.5 million tonnes consumed in the EU annually (Food and Agriculture Organisation 2006, p. 24). The threat to other ACP exporters is more immediate. As a result of its treaty obligations, the EU already imports more sugar than it requires. Consequently, if cuts were to be made, it is likely that they would be made against other non-LDC producers. The Caribbean sugar producers in particular are concerned about the repercussions of the EBA initiative, as exports to the EU constitute the bulk of their foreign earnings.

A similar picture materialises when the rice and banana sectors are considered. LDCs produce 41 million tonnes of rice annually, whilst consuming 43 million tonnes (Réseau Europeén de Sécurité Alimentaire 1999, p. 12). The fact that LDCs are net consumers of rice aside, the bulk of the rice produced by these countries would fail EU quality standards due to poor harvesting techniques. However, if LDCs were able to improve both the quantity and quality of their rice yields they could become a threat to non-LDC producers who are subject to tariffs. In the case of bananas, LDCs only produce 5.7 million tonnes, most of which are consumed domestically. Once again, the quality of LDC bananas is poor and as such would not constitute a major threat to other EU banana suppliers (Matthews and Gallezot 2006, p. 14).

EBA has also been criticised for merely being a 'catch-phrase'. That EBA is not enshrined in treaty form means that its provisions are not binding on the EU. If LDCs begin to profit too handsomely from the initiative, it would be possible for the EU to alter the terms of trade. Furthermore, EBA will only have a positive impact on LDCs if the goods that they export to the EU were originally subject to tariffs. As most LDC exports to the EU are already tariff-free as a result of the Cotonou Agreement, the impact is likely to be minimal. According to 1998 figures, the EU collected only €7 million ($9.1 million) from LDC goods entering the EU (Stevens and Kennan 2001, p. 3). World Bank figures estimate a boost of $185 million for LDCs as a result of EBA, an increase of approximately 1 per cent of income (Muyakwa 2001, p. 2). Up to 70 per cent of the potential benefits will only be fully realised by 2009, when all products besides arms are entitled to duty free access to EU markets. Furthermore, while the EU states that it hopes that EBA will boost and encourage production in LDCs, the strict rules regarding origin[2] make it difficult for these countries to expand their export bases. Processed foodstuffs have to be wholly produced in the exporting country whilst manufactured goods require additional value-added criteria, often at between 50 and 60 per cent of the total cost.[3]

The need for EBA

If EBA, as has been highlighted, is likely to have little significant effect on the economies of LDCs and is subject to caveats pertaining to sectors that might have bolstered LDC incomes, why then was the initiative implemented? There are three core areas of contention: that the EU introduced EBA just before the Doha meeting in order to curry favour with developing countries in the build-up to the conference; that EBA was an attempt by the EU to force the 'reform' of the Lomé regime and the destruction of the ACP grouping; that EBA was a tool which enabled liberalising elements within DG Trade to place the EU's sugar protocol under pressure. The motivation behind EBA is important as this is likely to have a direct impact on the initiatives' chances of success. Similarly, if the EU has adopted a cynical approach to its dealing with the poorest countries in the world, this would suggest that its dealings with the ACP bloc as a whole are likely to be similarly coloured.

On the face of it, the Doha hypothesis proffers a simple explanation. In the late 1990s, the EU was anxious to inject some vitality into the flagging WTO process. This was especially the case after the chaos and recriminations that followed in the wake of the Seattle conference in November 1999. EBA gave the EU the moral authority to call for the creation of a 'development round'. This was important for EU as it would, arguably, help to prevent a repeat of Seattle and facilitate increased global trade liberalisation. Furthermore, EBA 'cloaked' the EU's enthusiasm for a new round in almost altruistic terms. However, this line of argument is perhaps overly cynical. The LDCs carry little weight in the WTO and their ability to influence, obstruct or cajole is limited. In fact, as a result of the non-reciprocal elements contained within the initiative, it could be argued that EBA was just as likely to annoy the more powerful non-LDCs in the WTO, especially the so-called G-20 bloc. In this light, EBA seems an unlikely inducement with which to curry favour with other developing nations.

Others commentators have argued that EBA represents further evidence for the contention that the EU is determined to break up the ACP bloc. There has been a lingering suspicion in certain quarters that the EU was chafing under the restrictions of the Lomé regime (as outlined in Chapter 1) and was determined to remove the Bloc's influence on development policy. As stated, each renegotiation of Lomé resulted in a greater degree of dominance by the EU over its ACP partners. The sense of egalitarianism that prevailed during the initial negotiations in 1976 had all but dissipated by the time Lomé IV-bis came into effect. Cotonou's attempt to break the bloc up into regions and EBA 'special treatment' for LDCs further weakened the bloc bargaining power of the ACP countries. The EU's lack of sentimentality for the ACP as a group was made clear in its 1996 Green Paper pertaining to EU-ACP relations:

> The 'ACP group' is strictly speaking neither a political nor an economic entity. It was established in the framework of relations with

the European Union for essentially historical reasons. Today, the members of the ACP group have development interests and opportunities which diverge widely because of their very different income levels and living standards, their economic structures and trade strategies and their relative appeal for foreign investment. In this context, is the ACP group still an appropriate partner for the European Union? (European Commission 1996, p. 43)

In a sense EBA can be viewed as an extension of this logic. By including non-ACP countries in the initiative, the EU effectively destroyed the sanctity of the 'special' relationship between the EU and the ACP that had existed since the signing of the Georgetown Declaration in 1976, thereby allowing the EU to impose its vision on developing countries.

Moreover, it has been argued that the anticipation of a new WTO round served to radicalise DG Trade, resulting in it becoming a far more forceful player within the Commission. This in turn allowed DG Trade to press its neoliberal vision on anachronistic elements within the Polity. When the EBA restrictions on sugar are lifted in 2009 and the EU becomes obliged to allow the tariff-free, quota-free entry of LDC sugar into the Polity, the EU's Sugar Protocol will become difficult to sustain, particularly if imports increase significantly. It has been argued by Michael Brüntrup (2006) that EBA is but a 'Trojan Horse' for sugar reform. The tension between the 'reforming' mission of DG Trade and the 'stalling' attempts of vested interests represented by DG Agriculture were clear from the in-fighting and conflicting reports issued by the opposing parties. In November 2000, DG Agriculture published a report highlighting the dangers inherent in EBA to EU agricultural interests. This appraisal was far more pessimistic than that offered by DG Trade. The result was that, combined with pressure from the EU sugar sector the original EBA proposals were somewhat watered down (Obie 2007). However, the 'Trojan Horse' argument is an intriguing one, and one that offers a number of possible insights into the motives of EU policymakers in the run-up to the Doha Round negotiations and, by extension, their motivation in determining the direction of development policy.

The reason sugar is such an important element in the relationship between the EU and the ACP in general, and the EU-LDCs in particular, is that the EU sugar regime ensures that domestic prices are almost three times higher than the global average. The EU sugar sector remains one of the most regulated markets in the world, with high levels of intervention, price controls and subsidies. The regime allows for quotas, guaranteed prices and export subsidies. In order to protect domestic producers, it has ensured that imports struggle to compete on the domestic market by virtue of vast tariff barriers. Privileged access to EU markets can thus be extremely profitable for those concerned. Sugar is therefore an extremely 'value-added' commodity for those countries with access to the EU domestic market. Moreover, in terms of EBA, it is the one sector where, once fully liberalised, the LDCs have the ability to derive substantial benefits. Sugar is so important to the success of the EBA regime that the agreement has been dubbed, in some quarters, as 'nothing but sugar' (as opposed to its original dismissive moniker of 'everything but farms') (Cernat *et al.* 2003, p. 18).

Economic modelling suggests that Malawi is the best placed among LDCs to benefit from EBA and will achieve these gains by bolstering its sugar exports to the EU. The other main beneficiaries of EBA are likely to be Madagascar, Tanzania and Zambia, also largely as a result of increased sugar exports (Cernat *et al.* 2003, p. 18). It is thus clear that LDCs have a vested interest in seeing the sugar regime maintained.

Paradoxically, Brüntrup (2006, p. 13) argues that EBA and continuing pressure from the WTO[4] were largely responsible for inducing a significant reform of the EU's Sugar Market Order (SMO) in November 2005. EU agricultural ministers agreed to a number of reforms of the SMO, the most noteworthy being a staggered decrease in the guaranteed price for white sugar of 36 per cent, spread over four years, beginning in 2006. To compensate for this 'shock', EU producers would be well compensated, to the value of €6 billion ($7.8 billion) over the same four-year period (Brüntrup 2006, p. 14). What is notable about these sugar reforms was that the majority of actors involved were vehemently against the price cuts. The sugar producing LDCs actually offered to peg the level of their exports to 1.62 million tonnes per year in exchange for a continuation of the SMO regime until 2019. The Commission (2003d), however, refused. Moreover, the basis for its refusal was that it would run counter to EBA: 'one of the pillars of the agricultural proposal on market access in the WTO and other international fora', and that failure to reform the sugar regime would 'hurt the Community's credibility' (European Commission 2000c, p. 18). It can thus be argued, somewhat forcefully, that EBA was a tool for ensuring the forced reform of the sugar regime, which ran the risk of becoming an 'anomaly deviating from the fundamental principles of the new Common Agricultural Policy' (European Commission 2005b, p. 4).

EBA: altruism or opportunism?

What then are the benefits of EBA to the LDCs? Economic modelling appears to suggest that there will be welfare gains for these countries. However, these will be relatively small. Furthermore, the EU reserves the right to implement 'safeguard' measures if any LDCs exports flood the Polity's domestic markets. In order to ascertain the possible benefits of the initiative to the countries concerned, it is important to outline their pre-EBA position and then to consider the possibilities based on available economic projections. Of the $2.5 billion of agricultural goods exported by LDCs to the EU in 1999, only $100 million was subject to tariffs and restrictions, since phased out by EBA (Danish Research Institute of Food Economics, 2003). Nearly 70 per cent of this figure was tied to sugar exports, thereby underscoring the value of this sector to LDCs wishing to exploit EBA, as outlined above. Pre-EBA sugar tariffs stood at between 75 and 103 per cent (Yu and Jensen 2003, p. 9). Likewise, wheat, meats, rice, dairy and cereal grains all demonstrate relatively substantial tariff reductions and thus also represent key areas for LDC expansion. All in all, it is estimated that EBA will generate nearly $300 million in welfare gains for the LDCs but this is only if the three 'sensitive' sectors are opened up to them (Danish Research Institute of Food Economics, 2003). Malawi, one of the countries best set to exploit EBA would see its GDP rise by 4.7 per cent, Zambia by 2.4 per cent, Mozambique by 1.1 per cent and Sub-Saharan Africa by 0.5 per cent (Yu and Jensen 2003, p. 15). In monetary terms, this represents an increase of $30 million, $11 million, $8 million and $169 million respectively. These figures represent relatively modest increases and are somewhat undermined if the EU continues to reform its agricultural sector; as these gains will steadily diminish as a result of inevitably declining domestic prices. It is thus clear that EBA, despite its widespread publicity, represents little in the way of benefits for LDCs. The fact that the EU has used EBA as an excuse to reform its sugar sector in the face of domestic opposition and LDC requests, together with breaking the unity of the ACP group, suggests that perhaps this is what was intended for in the first place.

EBA vs. EPA

The LDCs are faced with a quandary. In essence, they could avoid becoming entangled with the EPA agreements and focus instead on bolstering their exports to the EU. On the surface this would appear to be the simplest option. It would mean that these countries would be spared the need to join a regional trading bloc, the benefits of which are likely to be negligible, and concentrate instead on exporting to the lucrative EU market. This approach, if enacted, would pose a number of problems for the EU's Cotonou strategy because a number of regions contain a statistically high proportion of LDCs and these blocs would be rendered unworkable if LDCs decided not to participate. However, LDCs are constrained from taking such a step by a number of issues. As stated, EBA is not enshrined in treaty form and could technically be withdrawn at any time. Furthermore, EBA has no aid component. The EU's development aid for ACP countries is tied to the Cotonou Agreement. These two factors make the EPA option difficult to ignore. While EBA gives LDCs the illusion of 'wriggle room' with regards to the EPA, this flexibility remains chiefly a facade.

Conclusion

The value of the EBA initiative to LDCs is thus difficult to gauge. It could theoretically have the potential to significantly boost LDC exports to the EU in coming years. If EU sugar interests are to be believed, member states could see themselves flooded by LDC sugar, rising from a current level of approximately 100, 000 tonnes per annum to a possible 4.3 million tonnes (with careful planning on the part of LDC producers). However, such increases are extremely unlikely and the net gains to LDCs from the initiative, whilst no doubt welcome, are likely to be relatively modest, given that the majority of LDC exports already entered the EU tariff-free before the implementation of EBA. Furthermore, whilst the initiative appears to provide the LDCs with increased options where their relationship with the EU is concerned, the reality is that EBA comes with no aid component, safeguards or guarantees and the options are thus largely illusionary. The EU, on the other hand, derived a great deal of positive publicity from EBA and has established its 'pro-poor' credentials in front of other major donors such as the USA and Japan. It can thus arguably be viewed as an inexpensive EU sop to developing countries in the build-up to the Doha Round. However, it has also been posited that EBA was not aimed at facilitating poverty alleviation at all, but was simply a Trojan horse designed to undermine the more anachronistic elements of the EU agricultural regime. What is indisputable is that there is a significant gap between the rhetoric of the EBA and the evidence in support of its benefits. Furthermore, the negative consequences for the ACP are quite severe. The initiative divides the ACP bloc into LDCs and non-LDCs. Moreover, a minority of LDCs are not actually ACP members, further undermining ACP unity – given that the EBA initiative ignores the 'special bond' between the EU and these countries. It has also generated a degree of antagonism between those ACP countries that qualify for EBA and those which do not. This is especially evident amongst the non-LDC Caribbean sugar producers. The proof of the pudding is in the eating and, as such, EBA should be judged, in conjunction with the Cotonou regime, on the benefits accrued to ACP countries with their 'gradual integration into the global economy'. These benefits will be assessed in Chapter 5.

5
Trade Policy, Poverty Alleviation and Sustainable Development

Introduction

Empirical evidence suggests that increased flows of trade generate wealth (O'Brien and Williams 2007, pp. 148–52). Given that trade issues form a core aspect of the Cotonou Agreement, it is important to evaluate the extent to which EU policies facilitate increases in trade flows between the Polity and the ACP and the extent to which these activities are sustainable in the long term. As highlighted, the rationale behind the Cotonou Treaty (Article 1) is poverty alleviation and the facilitation of the ACP countries into the global economy whilst simultaneously promoting environmentally sound outcomes. In this regard, trade policy clearly has an important role to play. The Cotonou Agreement remains in its early stages and a true assessment of its impact on sustainability is difficult to gauge before the implementation of the economic partnership agreements (EPAs) in 2008.

Being able to provide for basic subsistence is clearly the platform for any poverty alleviation strategy, and analysis of the 'poverty trap' provides a rationalisation for including environmental protection strategies within that platform. The EU bases its stance on the assumption that its neoliberal poverty alleviation strategies will result in both a cessation of environmental degradation and an increase in economic growth. This chapter challenges this assumption from three perspectives. First, it considers the evidence behind the equation of economic growth with lower levels of environmental degradation. Second, it evaluates the impact of neoliberal economic strategies on economic growth and poverty reduction. Finally, it discusses the extent to which wider environmental protection measures, while of benefit to the North, may be counterproductive where economic growth and poverty alleviation (and thus, by the EU's own assumption, localised environmental protection) in ACP countries are concerned. Poverty alleviation, far from evolving in tandem with environmental protection, is on the one hand endangered by it, in the form of environmental tariffs, and yet on the other still threatened by the lack of it, in that neoliberal economic strategies continue to emphasise environmentally unsound cash-cropping and the degradation on marginal land.

Trade policy, more than any other, illustrates the tensions inherent in any attempt to adhere to a sustainable development paradigm in as far as it can be taken as a simultaneous attempt to provide sound economic growth and poverty reduction by way of environmentally responsible means. As argued in the Introduction, the crux of the debate revolves around two basic positions: whether to promote policies centred on self-sufficiency or self-reliance when developing strategies for dealing with those living in poverty.[1] As discussed, self-sufficiency entails the ability to provide basic subsistence whereas self-reliance can be equated with the possession of a degree of purchasing power sufficient to acquire an adequate standard of living. Both positions involve a number of assumptions and risks. Self-sufficiency is often associated with heavy state involvement, such as that behind the 'green revolution' in India, in an era currently governed by an increasing 'roll back' of the state. Self-reliance strategies, on the other hand, are usually tied to neoliberal policies such as trade liberalisation and the participation of developing countries in the global economy.

Neoliberal theory dictates that trade liberalisation, the lowering of tariffs, a focus on exports and free trade are all central to the generation of wealth and the raising of living standards and thereby self-reliance. However, free trade and environmental considerations make uneasy bedfellows and many would argue that the two are diametrically opposed.[2] Despite rhetoric to the contrary, the prevailing trade and development orthodoxy, centred largely on neoliberal macroeconomic practices, and promulgated by all major international donors and financial institutions, the EU included, has become increasingly 'ideological' and inflexible since the advent of the 'Washington Consensus'.

A major difficulty when assessing EU-ACP trade relations is that the ACP is in no sense a homogenous bloc. It is geographically dispersed: consisting of microstates such as Nauru; densely populated states such as Nigeria; LDCs such as Mozambique; and political 'anomalies' such as Cuba. A one-size-fits-all policy is, by definition, almost bound to fail in such circumstances. The EU has acknowledged these variations and has put forward the 'Everything But Arms' (EBA) initiative as a response to the plight of the LDCs.[3] However, by splitting the ACP into two separate blocs, LDCs and non-LDCs, the EU has highlighted the myriad problems facing policymakers.

The two blocs have fundamentally different needs and demands. Such differences are clearly exposed when considering the effects of the EU's Common Agricultural Policy (CAP) on the economic fortunes of the ACP countries. There are recurrent calls from NGOs and campaigners for the EU to reform the CAP on the basis that is detrimental to farmers in the developing world who cannot compete with over-subsidised European farmers capable of overcoming their lack of competitive advantage in the agricultural sector through public funding. 'Trade not Aid' is the frequently made call, the assumption being that, by dismantling the CAP and laying a level playing field, low-income farmers will be able to compete with farmers from developed countries. Analysis of the 'Trade not Aid' position reveals that a dismantling of the CAP will indeed benefit some developing countries.[4] However, it is the countries from the so-called Cairns Group (Argentina, Australia, Bolivia, Brazil, Canada, Chile, Colombia, Costa Rica, Guatemala, Indonesia, Malaysia, New Zealand, Pakistan, Paraguay, the Philippines, South Africa, Thailand and Uruguay) which stand to benefit the most. The poorest countries in the world, the LDCs, would arguably actually be disadvantaged by the removal of these barriers to trade as a result of the EBA initiative, which permits LDC exports to enter the EU duty free thereby allowing these countries to sell their commodities at above world prices.

Furthermore, even if all barriers to trade were to be removed, a level playing field implemented and economic growth successfully fostered, a number of concerns remain. LDCs, even if they were to witness significant increases in per capita GDP, are likely to retain relatively high poverty levels for the foreseeable future.[5] This in turn emphasises the fact that for many in these countries, the natural environment will remain a primary source of subsistence. It is thus important to assess whether pro-trade policies will put additional strain on often already fragile natural systems and therefore, in the medium term at least, possibly leave people worse off.

A hidden trade agenda?

Some critics view the Cotonou process as an attempt by the EU to achieve advantages it has been unable to gain where the WTO is concerned (Griffith and Powell 2007). As outlined in Chapter 3, developing countries have succeeded in asserting themselves within the WTO. The formation of the G-20 and G-90 has resulted in increased cooperation and a commonality of cause amongst countries in the South. The upshot has been that progress within the WTO has ground to a halt in the face of a refusal by developing countries to sanction new measures on investment, public procurement and competition policy – the so-called 'Singapore Issues' – until their demands pertaining to the development aspects of the Doha Round are met.

The EPAs (see Chapter 9) demanded by the EU as part of the Cotonou Agreement are in effect 'WTO-plus' in that they go further than the current agenda being advanced at the WTO – the agenda already rejected by the majority of developing countries. This strategy arguably represents trade liberalisation through the 'back door' and is, for some, the true motivation for the new regime. The Commission (2006c, p. 2) itself acknowledges that its EPA approach will increase the international competitiveness of its exports. However, it can also be posited that since EU policymakers maintain that trade is a driver of economic growth, demands for greater liberalisation are perfectly inline with the 'spirit of Cotonou':

> Through our trade policies, we also seek to contribute to a range of the Union's external goals, in particular development and neighbourhood objectives. These will remain core functions of EU trade policy. Coherence of the Union's external policies is vital to strengthening the EU's global role... [W]e will need to work to strengthen sustainable development through our bilateral trade relations. (European Commission 2006c, pp. 2, 9)

The role of trade as a vehicle for promoting sustainable development needs to be gauged in order to assess its worth as a strategy for ACP country development. NGOs such as Oxfam, ActionAid, CAFOD, Traidcraft and Christian Aid have all seriously questioned the EU's agenda, both in terms of motive and veracity. They argue strongly that 'trade liberalisation should not be seen as a substitute for a sound development strategy' (Stop EPAs Coalition 2004, p. 2). This scepticism is seemingly warranted when the potentially conflicting goals of the Lisbon Agenda and Göteborg Agenda are considered. The Lisbon Agenda, adopted at the European Summit in Lisbon in 2000, had two clear goals – transforming the slowing EU economy into the 'most competitive economy in the world' and providing full employment within the Polity by 2010 (European Council 2000). The EU soon came under public pressure to include an 'environmental pillar' to complement the economic and social elements explicit within Lisbon. Accordingly, the 2001 Göteborg European Council Summit added the requisite environmental emphasis. The official statement released by the Council (2001, p. 5) stated that '[i]n particular, the EU should promote issues of global environmental governance and ensure that trade and environment policies are mutually supportive'. The extent to which the EU has been able to afford equal priority to each of these pillars is debateable and, for many, the sustainable development component is simply an 'add-on' to appease environmental interests. Furthermore, even assuming that a demand for greater liberalisation on the part of the ACP countries represents a genuine development strategy rather than an 'offensive interest', to what extent is trade-induced sustainable development likely? An independent study commissioned by the EU states

> [L]iberalisation, if not accompanied by considerably strengthened supporting measures, is likely to result in a much sharper division between countries and socio-economic groups into gainers and losers in economic welfare, particularly during the short and medium term adjustment period. In the longer term, assuming markets become more efficient, additional economic welfare gains should be made by these country groups as a whole. However, it is more problematic to establish the extent to which more countries and social groups will share in these welfare gains or, more fundamentally, how many of these countries will emerge from the transitional process on a closer trajectory to sustainability than previously. (Kirkpatrick and Lee 1999, p. 117)

Trade and the environment

Jacques Morisset, the World Bank Project Manager for Africa, postulates that, for countries with few valuable natural resources, the most successful route to development is trade liberalisation. By the late 1990s, both Mali and Mozambique had instituted economic 'reforms' in an attempt to create investor confidence and as a result had achieved growth rates of 7 per cent and 13.3 per cent respectively by the end of the decade. This growth was achieved with single figure inflation (Morisset 2000, p. 17). Both countries were able to attract a larger share of foreign direct investment (FDI) than Kenya and Cameroon, despite the latter having larger domestic markets and greater access to natural resources. This, Morisset claims, demonstrates that, by merely creating the right economic conditions, countries can encourage good rates of economic growth.

The theory that people are forced to deplete and degrade their natural surroundings in order to survive invites the conclusion, on the part of pro-liberalisation economists, that increased attention to poverty alleviation will promote a corresponding decrease in environmental degradation. Empirical evidence does tend to suggest a link between levels of GDP and environmental protection. The countries which demonstrate the highest levels of GDP also maintain the most stringent environmental legislation (Esty and Levy 2006). Thus, empirical evidence suggesting that liberalisation may drive economic growth (Morrison and Pearce 2003, pp. 3–24) forms the basis for claims that increased trade will result in decreased poverty levels and thus greater environmental protection.

However, for the EU to claim success in policy areas relating to both poverty alleviation and sustainable development, the evidence linking economic growth and environmentally sound outcomes must be clear. It is not. In the 1950s, Simon Kuznets plotted the relationship between per capita income and economic growth. His results suggested that, as an economy begins to develop, inequalities in income initially widen, peak and then fall. Plotted on a graph, these figures created an inverted 'U' shape – an outline commonly described as a 'Kuznets Curve'. In the early 1990s it was demonstrated that increased per capita income and pollution indicators produced a similar curve when plotted on a graph (Panayotou 1993; Grossman and Krueger 1994; Shafik and Bandyopadhyay 1992; World Bank 1992) The logical conclusion was that developing economies should accept environmental degradation as part of the development process on the basis that this would be a temporary phase. Water and localised air pollution in particular demonstrate a correlation between economic development and increased environmental standards. The US Federal Reserve Chairman, Alan Greenspan, testifying before a Senate Finance Committee in April 2001, stated forcefully that as far as he was concerned, 'environmental quality [is] directly related to the degree of prosperity in a particular economy' (cited in Griswold 2001, p. 24N). However, this is not always the case. Unfortunately for proponents of the 'growth approach', few transboundary emissions, such as 'greenhouse' gases, decrease in the face of economic prosperity. In fact, the reverse is true, as the countries most responsible for these emissions are the developed nations. Robert Ayres

(1995, p. 97) describes the notion that economic growth is 'good for the environment' as 'false and pernicious nonsense'. He argues that part of the reason that pollution levels decline in richer countries is because 'dirty' industries relocate to countries with cheaper labour and lower standards. Furthermore, he argues that economic growth is almost always closely linked to an increase in energy consumption, invariably extracted from unsustainable fossil fuel sources and thus in no sense 'environmentally friendly' (Ayres 1995, p. 97). A 2007 sustainability impact assessment conducted into the possible effects of EPAs argues that whether 'development occurs and whether or not it is sustainable, depends not only on trade, but on a host of other factors' (Krimphoff 2007, p. 16). Simply bolstering trade flows through greater liberalisation is insufficient. Nonetheless, an Organisation for Economic Cooperation and Development (OECD 1998, p. 20) study concluded that 'There is some evidence that, once a country begins to industrialise, trade liberalisation helps to make the structure of its economy less pollution-intensive than in those countries which remain relatively closed.'

Since few LDCs are beginning to industrialise in any meaningful sense, such benefits are likely to apply less to them than developing countries in the Cairns Group. However, it can be argued that increased per capita income leads to less immediate reliance on natural surroundings for subsistence (self-reliance) and thus can have a positive environmental impact on the patterns of degradation usually associated with poverty. Then significantly increased levels of trade between the EU and the lower-income ACP countries will go a long way towards fulfilling the EU's desire for an economic system based on the principles of sustainable development and poverty alleviation.

However, questions remain. Even if it were satisfactorily demonstrated that increased per capita income results in higher environmental standards, it is far from clear that policies designed by the EU to promote such development amongst its ACP partners will be successful in this regard. The failure of this strategy could potentially both render self-reliance on the part of these countries unlikely, and undermine any possibility of self-sufficiency.

The North–South divide

The EU is not alone in its call for greater liberalisation on the part of the ACP countries and represents part of a wider consensus. The World Trade Organisation (WTO), International Monetary Fund (IMF), and World Bank all actively encourage developing countries to reduce subsidies and tariffs in order to make themselves more competitive and to promote export-led growth. However, the developed countries which dominate these bodies do not necessarily practise what they preach. The EU's CAP is once again at the fore, regularly cited as being in flagrant breach of all neoliberal principles, dispensing approximately €55 billion ($71.5 billion) per year in subsidies (European Commission 2006d). The USA and Japan, while not as generous, also heavily subsidise their agricultural sectors (Hoekman and Kostecki 2001, p. 209). These subsidies have a negative impact on free trade. For example, the CAP regularly results in large-scale overproduction. The excess is then frequently 'dumped' on developing countries, thereby undercutting local producers who cannot compete. Countries in southern Africa, having liberalised their agricultural sectors, have in the past been swamped with EU-subsidised beef and tomatoes, which in turn have marginalised local producers. Similarly, dairy producers in the Caribbean have been forced to compete with heavily subsidised milk powder emanating from the EU (Godfrey 2002, pp. 5–6). A study by the Food and Agriculture Organisation (FAO) has found that, since the Uruguay Round, which stipulated a 20 per cent decrease in agricultural subsidies by 2001, food imports have flooded into developing countries, while exports from the latter have shown little sign of improvement (cited in Madeley 2000, pp. 71–3; see also Kipe 2003).

Furthermore, while access to the markets of developed countries now appear, at first glance, to be relatively straightforward, significant barriers to trade still exist. While EU tariffs since 1999 have averaged just 7 per cent, it is important to note which sectors contain the 'peak' tariffs. Some of the highest barriers to goods entering the EU are to be found on agricultural produce. Cereals, bananas, sugar and meat all face stiff tariffs when entering the EU – some as high as 100 per cent (Hoekman *et al.* 2001, p. 1). Likewise, the fishing industry is estimated by the World Bank to be subsidised to the sum of $20 billion worldwide (World Bank 2005a). Figures suggest that up to 25 per cent of fishing revenues now consist of subsidies, the vast majority of which are issued by developed countries (OCEANA 2006). It is claimed that 90 per cent of these subsidies are in violation of WTO rules (Insausti 2001). This represents a potentially disproportionate impact on developing countries, many of which rely on agriculture or fisheries for exports. Thus, even if free trade is a panacea for environmental degradation in developing countries, it is clear that the global economic system will have to be far more open if this is to occur. Yet, despite the promises made at the conclusion of the Uruguay Round, subsidies in developed countries are continuing to rise. Between 1995 and 2004, agricultural subsidies for the OECD countries increased from $182 billion to $300 billion (La Vina *et al.* 2006).

Mountains out of molehills?

It is argued in some quarters, however, that such statistics are largely 'smoke and mirrors' and do not fully represent the realities of international trade. Arvind Panagariya (2005, 2006), at various times advisor to the World Bank, IMF and WTO, argues forcefully that the impact of many of these supposed barriers to trade have been greatly exaggerated. He is concerned that there is bourgeoning dogma, beginning to verge on orthodoxy, which lays all the blame for the developing world's ills at the door of developed countries. Oxfam, for example, argues that if rich countries would liberalise their markets and eradicate subsidies, it would be worth upwards of $100 billion in additional revenue for developing countries (Watkins and Fowler 2004). The famous UNDP statistic exposing the EU's expenditure of more than $913 in subsidies on every cow in the Polity juxtaposed against a seemingly paltry $8 per person in aid to Sub-Saharan Africa is another frequently cited piece of evidence used to demonstrate the inequities inherent in the CAP (cited in Panagariya 2006). Panagariya maintains that such comparisons are spurious; all countries spend comparatively more addressing internal issues than they do on international aid; and that this is equally true for developing countries. Moreover, the effects of export subsidies, long pilloried by NGOs and poverty campaigners, are, in his view, overstated and increasingly irrelevant. Export subsidies, based on WTO figures, suggest that such inputs are now in the vicinity of $3–5 billion. The elimination of such subsidies will, in Panagariya's (2005) view, provide little overall benefit for the international trading regime. Once again, extrapolating on WTO figures, he attributes a figure of $100.7 billion in subsidies for the top five domestic subsidy users, including permissible blue-box and *de minimis* subsidies. Whilst undoubtedly high, this figure is substantially lower than has been suggested. Former World Bank President Paul Wolfowitz, for example, has put the figure at $280 billion, while Oxfam has estimated it to be in excess of $300 billion (cited in Panagariya 2005). Such starkly differing totals have arisen as a result of analysts employing different criteria when defining subsidies. The lower figure is based on trade-distorting export subsidies and amber-box subsidies whilst the higher estimates are based on the OECD's Producer Support Estimates (PSE) which measure the total income in excess of the global market level, be it on the basis of WTO-sanctioned interventions or not. Panagariya asserts that the higher figures proffered would be disputed by the majority of economists as many such 'subsidies' are not in violation of WTO rules. Consequently, he asserts that although these figures are not trifling, the benefits attached to the removal of subsidies are lower than many would suggest. Tariffs, Panagariya claims, are more significant where the curtailment of the flow of goods is concerned.

EU 'Green' trade barriers

For many campaigners, there is a fear that, with mounting pressure to reform the international trading regime, wealthy countries may resort to new forms of protectionism. As traditional trade barriers are slowly and steadfastly eroded, even in the face of obstructionism, there is a fear on the part of developing countries that wealthy states might erect non-trade barriers in their stead in order to check what might be viewed as excessive imports. Many developing countries are concerned that environmental and technical requirements might replace tariffs and duties. 'Green' tariffs are a particularly sensitive issue. The use of such tariffs has provoked a storm of controversy and is opposed by free trade economists and development campaigners alike. Many in the South believe that developed countries are attempting to use environmental standards as a pretext for maintaining, if not strengthening, trade barriers.

There is special concern regarding the Application of Sanitary and Phytosanitary Measures (SPS) which forms part of the Uruguay Round Agreement. The SPS Agreement allows countries to block imports of goods that either represent a threat to human or animal well-being or pose a significant risk to the environment. Exporters must be able to demonstrate a variety of checks including information on traceability, pesticide residue levels and the use of additives. Currently, SPS legislation represents one of the greatest stumbling blocks to ACP goods competing on EU markets. It must be stressed that SPS requirements are important. They allow importing countries to protect themselves from pests, threats to public health and environmental hazards. These policies are in place to halt, amongst other things, the spread of diseases such as the so-called 'zoonoses'; diseases which can be transmitted from animals to humans, and which include tuberculosis, salmonellosis and listeriosis.[6] The zoonoses represent a serious threat to human health. Listeriosis, commonly found in un-pasteurised milk, can cause encephalitis and spontaneous abortions. Nevertheless, there is widespread concern amongst developing countries that SPS measures might increasingly be used as a form of tariff. Most SPS legislation is aimed at agricultural products and, as the majority of ACP exports constitutes such goods, these countries stand to be disproportionately affected. In the wake of the BSE crisis, the genetically modified (GM) food debate and foot-and-mouth, the EU has imposed stringent food safety measures. It requires procedures that will accommodate a 'farm to table' system of checks and controls. The aim is to ensure the regulation of all aspects of food production, including the manufacturing of feed, processing and storage (European Commission 2000d, p. 8). Another 'principle of food safety' entails the traceability of all food and feed. Moreover, third countries must be able to guarantee that all products exported to the Polity conform to EU safety standards (or equivalent) (Doherty 2006). This means that detailed records relating to suppliers and customers must be maintained, with regular tests confirming quality control.

Intentionally or not, SPS measures represent a trade barrier to developing countries, especially LDCs, and threaten to negate any positive aspects made in dismantling other trade barriers such as export subsidies and traditional border tariffs. The standards demanded by the EU's food safety proposals entail high levels of administration and an extensive testing/monitoring capacity. It can sometimes take between two and four years to demonstrate that a country's produce meets the necessary requirements, but since the EU does not publish figures on the extent to which imports fail SPS measures, it is sometimes difficult to assess the impact of this legislation on ACP products. However, US data are published, and from these a general appreciation of the position of ACP countries wishing to export to the EU can be gleaned. Vegetables, fish products and fruit are the items most commonly rejected on the basis of their failure to meet SPS criteria (Unnevehr 2000, pp. 231–40). During the past 20 years, most were detained due to rodent or insect infestation. Other common violations include microbiological contamination and high pesticide residues.

Between 1984 and 1994 Guatemala had 3000 shipments of fresh produce blocked by American inspectors as a result of overly high pesticide residues. Consequently, the American Food and Drug Administration (FDA) now automatically detains all Guatemalan products entering the USA for testing. In 1996, Bangladesh exported $288 million in fishery products to the USA, Europe and Japan. Inspection of the processing plants by EU monitors in 1997 led to a ban on such products entering the EU as a result of poor hygiene practices (Unnevehr 2000, pp. 235–36). The Bangladeshi

Government had to invest substantially in order to restore confidence in these plants.

For many African countries, a major source of concern is the institution by the EU of increased standards relating to aflatoxins. These toxins are carcinogenic chemicals produced by the *Aspergillus* mould, which is commonly found on groundnuts (International Trade Forum 1999, pp. 8–9; Otsuki *et al.* 2001, p. 266). Contamination can occur at any number of stages of production and as a result strict control measures are needed to ensure containment. The EU use of SPS measures has increased this burden. By raising the guidelines governing aflatoxins from two parts per billion (ppb), as per the generally accepted international standard, to nine ppb, the EU puts at risk $670 million worth of groundnut exports from African countries (Muyakwa 2001, p. 4). Many lower-income countries simply cannot afford to implement such strict monitoring policies and as a result are unable to meet the required standards. The issue of SPS measures is likely to constitute a continuing problem for developing countries as their implementation is on the increase. The WTO data show that in 1995 there were 220 SPS notifications (Wilson 2001, p. 6). By 2004, the number of notifications had risen to 5,240, the vast majority of which were made by developed countries (World Trade Organisation 2004).

A similar pattern emerges regarding disputes referring to the SPS Agreement. Since the completion of the Uruguay Round, 32 complaints have been made, with most of these stemming from developed countries (WTO 2007, p. 5). The USA leads the way, followed by Canada and the EU. Only two developing countries, India and Thailand, have made similar complaints. No LDCs have attempted to force a ruling on SPS measures. This is partly due to the fact that, from a low-income country's perspective, there is little to be gained from WTO proceedings. If the Dispute Body rules in favour of the complaining party, then that country can impose punitive trade tariffs against the offending party. However, for the majority of developing countries, such sanctions could be counterproductive and would be more likely to harm their own economies.

Comparative advantage, trade regimes and poverty alleviation

Despite these obstacles, low-income countries are being encouraged, by the EU and other donor bodies, to exploit their comparative advantage in the agricultural sector by planting cash crops. Such programmes inevitably lead to some degree of mono-cropping. From an environmental perspective, mono-cropping and genetic conformity incorporate a number of risks, including susceptibility to specific plant diseases capable of eradicating whole harvests. The banana industry has a long history of such disasters. In the early part of the last century, subsistence farmers in Central America and the Caribbean tended to grow a variety of bananas. Demand for uniform size and colour led plantation owners to insist on mono-cropping. These plantations rarely lasted more than eight to ten years before they were decimated by disease (Tucker 2000).[7] Moreover, banana and sugar plantations have historically resulted in the large-scale clearance of tropical forests in LDCs such as Haiti, as well as the abundant use of pesticides and artificial fertilisers. A similar case can be made for coffee and cocoa production.[8]

Bolstered levels of trade are supposed to encourage developing countries to diversify and to find ways of adding value to their exports and yet, in reality, current development practises discourage multiformity, thereby perpetuating the reliance on low added-value agricultural produce. A sustainability impact assessment of the Cotonou regime, commissioned by the EU, highlights this very issue (Krimphoff 2007, pp. 16–17). A lack of diversification is problematic when making the case for liberalisation, as the terms of trade for agricultural produce have been steadily declining. Here, the coffee industry is a useful example. Despite a booming coffee market in developed countries, coffee prices in 2002 were less than a third of what they were in 1997 (Morrissey 2003, p. 43). In 1993, the global coffee market was estimated to be worth $30 billion, of which the producers' share was $12 billion. By 2003, the global coffee market had grown to $50 billion yet the share claimed by producers had contracted to $8 billion (Murphy 2003, p. 124). This pattern has been replicated across a number of agricultural sectors, which saw the terms of trade for countries in SSA decline by 9 per cent in the late 1990s. Unless agricultural produce can have value added, these commodities are unlikely to bring about the type of economic gains low-income countries such as the LDCs require in order to becoming self-reliant. Critically, EU legislation creates a number of obstacles to exports from developing countries. Tomato paste, canned peaches, canned pears, citrus fruit juice, wine and spirits, butter and milk powder are all heavily subsided or protected by EU legislation, thereby making it difficult for countries to break into these markets (Stuart 2005). Furthermore, the EU's SPS legislation makes value-added food production onerous, as processed food products need to be packaged in a specific manner in order to meet health and hygiene specifications. Most LDCs do not have the capacity to produce these packaging materials, resulting in the need to import them. The associated costs are prohibitive for low-income countries.

A further obstacle to ACP exports is the EU's complex and unwieldy 'point-of-origin' rules. Such legislation outlines the stipulations that must be satisfied in order to ascertain the country of origin of any given product. While point-of-origin measures are necessary in order to prevent 'trade deflection', whereby goods are re-directed through countries holding preferential trade agreements with the EU, they present a problem for a number of ACP exporters (Brenton and Manchin 2002, p. 2). They can represent a significant barrier to exporting countries, especially when they are accompanied by stringent technical regulations. It is a relatively simple procedure to gauge the origin of items that are single stage productions, such as agricultural goods. In other instances, complexities arise. The higher the level of manufacture behind the product, the more time-consuming it is to ascertain the origin of a commodity. There are myriad costs associated with detailing origin and the process requires a certain administrative capacity. The situation can become more complex if the countries involved have lax customs controls. If documentation is poor or lacking, it might be impossible to satisfy the necessary requirements. It has been found that only one-third of imports from developing countries actually meet the requirements enabling them to qualify for the preferential treatment to which they are entitled. This represents a significant under-utilisation of EU preferential schemes which, theoretically, offer preferences for 99 per cent of exports from developing countries (Brenton and Manchin 2002, pp. 1, 7). This is particularly the case where clothing and textile imports are concerned. The low cost of labour in many developing countries is what makes them attractive to clothing and textile producers, many of which have relocated to developing countries for this very reason. However, the annexes governing technical and point-of-origin regulations for clothing and textile manufacturing run to over 80 pages. Few ACP LDCs have the technical or administrative expertise to adequately deal with such regulations.

In order to qualify for aid and other forms of assistance, developing countries are being expected to institute economic reforms and open up their economies to foreign direct investment. A FAO study has shown that the liberalisation of agriculture in developing countries tends to lead to a concentration of land ownership as well as large-scale, plantation-style farming methods. It has also been argued that it results in the widespread displacement of rural populations (cited in Madeley 2000, pp. 73–9). As low-income countries such as the ACP LDCs often have extremely underdeveloped manufacturing sectors, there is little hope that these displaced people will find employment within other sectors of the economy. Consequently, they are forced either to farm marginal lands or to clear new areas for cultivation. This has widespread environmental ramifications, including deforestation and desertification (Tucker 2000, pp. 60–2).

The EU's demand for liberalisation is problematic in another sense. The economic theories upon which neoliberal strategies are based assume functioning markets. There is little sense in preparing developing countries for participation in the global market place if the market forces necessary to ensure the functioning of such systems are not in place. If these market conditions are not satisfied, then the full implementation of the Uruguay Round Agreement on Agriculture will make little difference to many lower-income ACP countries. This is an aspect of economic reform that is seldom discussed, but what is clear is that the international trade in primary commodities is becoming increasingly dominated by a relatively small number of multinational corporations (MNCs).

When the dominant four firms in a given sector account for more than 40 per cent of the market share, many economists argue that effective competition becomes difficult (Murphy 2003, p. 121). Yet this is exactly what is being replicated throughout the sectors in which developing countries hold a competitive advantage. To draw upon the coffee market once more, nearly 90 per cent of global coffee exports are handled by just three companies. Such clear dominance enables companies to ensure that the coffee prices paid to producers remain low, with the result that farmers need not necessarily benefit from increased production or volume of sales. Such dominance occurs because the commodity markets are fraught with risk and only large companies can afford to 'ride out' fluctuations in the global trading system. The unpredictability of commodity markets makes them too volatile for smaller ventures. The MNCs tend to absorb this risk by expanding both horizontally (for example, buying, shipping and milling grain) and vertically (for example, owning joint ventures that allow a company to participate in numerous aspects of the supply chain, such as both the growing and canning of fruit). By so doing MNCs can counter the perils inherent in changeable markets and absorb the costs associated with any potential shocks.

Evidence of further deterioration in operating markets is reflected in the trend towards the expansion of supermarket chains into developing countries, even in African states where, at first glance, there would appear to be little outlet for such ventures. Shoprite Checkers, a South African based multinational, is one that has taken advantage of increasingly open African markets in order to penetrate neighbouring countries. The company now has outlets in over 16 African countries, including LDCs such as Tanzania and Lesotho (Shoprite Holdings Ltd 2007). However, evidence from Latin America suggests that when such penetration occurs, a relatively small number of companies rapidly achieve dominance. Statistics suggest that up to 60 per cent of the supermarket sector in Latin America is controlled by between four and five firms (Reardon 2003, p. 132). The result is that certain commodity sectors have found themselves beholden to supermarket giants intent on cutting their costs. In Latin America, more than half of all dairy products are now sold through supermarkets, as are an increasing percentage of local fruit and vegetables. Not only are supermarket chains able to use their buying power to dominate market share but they also tend to bring with them 'First World' measures such as demands for levels of hygiene, packaging, fruit/vegetable size, shape and variety. In essence, international SPS measures are being imposed on domestic markets and producers. Furthermore, by shifting away from wholesalers to contractors, these supermarket chains are eliminating any bargaining power producers might have held in a functional market place.

The associated costs of globalisation

Even accepting – for a moment – that economic growth leads to environmental protection by virtue of reducing poverty, it is difficult to see how the EU's development policies will have the necessary effect in this regard. What is certain is that many of the policies will lead to added environmental pressures as people struggle to cope with the impact of trade liberalisation on employment and on the agricultural sector. The costs associated with many developing countries becoming functioning members of the international economy are – comparatively – vast. The reorientation of their economies involves dramatic bureaucratic and administrative changes, as well as demands for an increased scientific and technical capacity. Furthermore, in order to ensure that they are capable of 'playing the system', the lower-income members of the ACP bloc need greater representation within international bodies such as the WTO. That, as a result of the cost, nine LDCs have no representation in Geneva, is a clear indication of their continuing marginalisation

(United Nations Development Programme 2007, p. 12). The lowering of trade barriers alone will not suffice.

As yet there is no clear strategy in place aimed at aiding ACP countries in their efforts to make these adjustments. Where attempts have been made to lessen the financial blow, they have tended to be inadequate. In 1999, the EU and South Africa concluded a free trade treaty, the Trade, Development and Cooperation Agreement (TCDA), which committed both parties to a substantial lowering of trade barriers within the coming decade.[9] However, by doing so, South Africa, as the dominant member of the Southern African Customs Union (SACU), *de facto* committed its fellow members to the free trade agreement. This is significant. The EU is southern Africa's largest trading partner and hence the loss of tariff revenues are certain to be notable. To put the effects of the TDCA into perspective, Swaziland has traditionally derived nearly 50 per cent of its revenues from the SACU. Estimates suggest that the losses associated with the implementation of the TDCA are likely to be in the region of 13–14 per cent of Swaziland's annual income (approximately R420 million/$57 million – based on data provided by Kirk and Stern 2003, p. 5). The EU agreed for a sum of $1.3 million to Swaziland as compensation (Goodison and Stoneman 2005, p. 23).

As stated previously, EU interaction with its ACP partners cannot be viewed, or for that matter judged, in isolation. Consensus on the part of the majority of aid donors, including the EU, has resulted in a largely neoliberal approach to development, encompassing economic liberalisation, the removal of trade barriers and a 'rolling-back' of the state. While there is empirical evidence to suggest that such steps do indeed foster economic growth, there are concerns pertaining to both poverty alleviation and sustainable development outcomes. The majority of the world's poor live on marginal land in rural areas, and are heavily dependent on their natural surroundings in order to subsist. The degradation of the natural environment potentially could have a profound effect on their already precarious life chances. The neoliberal model assumes a boat lifted by a rising tide, sweeping all along. However, it has been argued that the benefits of even sustained, high level growth need not necessarily filter downwards. Moreover, the neoliberal model can often be accompanied by negative environmental consequences such as increased pressure on arable land. In many instances, neoliberal programmes have been accompanied by higher levels of unemployment together with a significant proportion of rural people being forced off their land in the face of expanding commercial agriculture. This in turn shifts people onto environmentally vulnerable land, which is frequently at risk of degradation. It is thus crucial that, if the EU wishes to persist with its current approach to development, sustainability concerns and self-sufficiency strategies be afforded far greater weight when policy is formulated.

Conclusion

In tying poverty alleviation and sustainable development together in the Cotonou Agreement, the EU has succeeded in formulating a development strategy that appears to 'tick all the boxes', being both pro-poor and 'environmentally friendly' whilst conforming to the rules of the global trading system. In doing so, however, it has created a rod for its own back, since, in viewing these elements as being almost synonymous, it ignores a number of conflicting tensions. The majority of the world's poor live on marginal land in rural areas, and are heavily dependent on their natural surroundings in order to subsist. The degradation of the natural environment potentially could have a profound effect on their already precarious life chances. However, there is little evidence that the growth-orientated development model being pursued by the EU will protect or preserve these fragile environments, thereby undermining the possibility of food self-sufficiency. Nor is there evidence to suggest that EU policies will affect a degree of wealth sufficient to render this problem void. Even when affording the EU policymakers the benefit of the doubt, it is difficult not to argue that policies such as the CAP, SPS legislation and rules of origin, together with the ACP's loss of tariff revenues, seriously undermine the EU's own stated development agenda. Returning to the three stated aims of the Cotonou Agreement, 'poverty eradication, sustainable development and the gradual integration of the ACP countries into the world economy', it is arguable that only the third appears to be realistically achievable through the implementation of the EU's current neoliberal policies.

6
Agriculture, Food Security and Rural Development

Introduction

Globally, nearly 70 per cent of people living in extreme poverty live in rural areas (European Commission 2000a, p. 5). If the Millennium Development Goal (MDG) of halving poverty by 2015 is to be achieved, then special attention needs to be devoted to rural economies, and specifically to agricultural policies. While the majority of the poor in the Caribbean[1] and Pacific[2] ACP countries also live in rural areas, the severity of the problem is most pronounced in Sub-Saharan Africa (SSA). In global terms, poverty is on the decline, with the number of people living in absolute poverty being set to fall from 1.27 billion in 1990 to 0.75 billion by 2015 (Broca *et al.* 2003, p. 216). However, these global averages belie regional imbalances. East Asia and South Asia have made significant progress over and above their 1990 MDG baseline figures and appear set to continue in this vein. The situation in Africa, on the other hand, is set to worsen. Since 1990, levels of poverty in SSA have increased and, it has been argued, are likely to increase even beyond 2015. If these projections are correct, then by 2015 nearly half of all those living in absolute poverty will be resident in Africa (Broca *et al.* 2003, p. 216). As the vast majority of Africans living in absolute poverty dwell in rural areas, it is imperative that rural development programmes are successful in targeting hunger and malnutrition, boosting the agrarian economy and ensuring environmental protection. Success in these areas will serve the dual purpose of increasing both welfare and economic growth. In LDCs, agriculture is responsible, on average, for 40 per cent of GDP and approximately 33 per cent of exports, whilst generating 70 per cent of all employment (European Commission 2000a, p. 6). A detailed, systematic approach to combating rural poverty is therefore a prerequisite for combating poverty in the ACP bloc as a whole, given that studies suggest that countries that have achieved significant poverty alleviation amongst their populations have done so on the back of developments in their agricultural sectors. Up to 70 per cent of these reductions in poverty can be attributed to growth in the area of agriculture and such growth has been shown to increase agricultural and non-agricultural employment, reduce poverty and improve basic development indicators (World Bank 2006a, p. 11–12).

The first goal for any development programme, especially in rural areas, must be food security, as hunger and poverty are inextricably linked. During the period 1961–63, developing countries generated, on balance, a $6.7 billion surplus in food exports. By the end of the 1990s this surplus was negligible, and by 2015 developing countries are likely to be net importers of food. A 2003 Food and Agriculture Organisation (FAO) publication estimates that this deficit will total $61 billion by 2030 (in 1997/1999 dollar values) (Schmidhuber and Raney 2003, pp. 235, 239). When one considers the relatively few 'success stories' involving developing countries that have attained food self-sufficiency and burgeoning agricultural exports, these have generally been achieved at a significant environmental cost.

From analysis of government-sponsored 'green revolutions' in Asia and Latin America since the 1960s, it is clear that many of these were successful in achieving bolstered levels of agricultural production. Following the Second World War, there was a growing concern that food shortages in developing countries could result in the outbreak of catastrophic famines. In response to this concern, a number of 'miracle' seeds were developed in the 1950s, which, in tests, produced higher than average yields. This new seed technology focused mainly on cereal crops such as rice, wheat and maize and was heavily promoted by the development agencies from the 1960s onwards. Countries such as India were amongst the first to adopt the new technology and the initial results were encouraging. However, a number of difficulties began to present themselves. The genetic base for these new seeds was very narrow and as a result these varieties were highly susceptible to inroads by pests and disease (Shiva 1991, p. 81). Consequently, it became necessary for scientists to develop new, hardier strains on a regular basis. In addition, in the Indian experience, the eradication of crop rotation and other traditional practices has led to the depletion of the soil and a dependency on artificial fertilisers. Nevertheless, it can be accepted that these 'green revolutions' have succeed in reducing poverty levels and boosting rural incomes. Furthermore, it can be demonstrated that increased levels of agricultural productivity in these regions have resulted in less land being cleared for farming.

On the face of it, then, the results speak for themselves. Over the past four decades per capita food production in Asia has increased by 40 per cent whilst Africa, which has not experienced a green revolution, has witnessed a 20 per cent decline over the same period (European Commission 2000a, p. 26). From an environmental perspective, however, these gains were made possible only through significant agrochemical input in the form of, for instance, fertilisers and pesticides. Furthermore, green revolutions are invariably associated with practices such as mono-cropping and large-scale irrigation schemes which have, in many cases, led to salination of the soil. India is a notable example in this regard. Moreover, economists have argued that self-sufficiency is often costly and inefficient, India once again being an example (Ziegler 2006, pp. 12–13).

The above would seem, therefore, to suggest a tension between the achievement of food self-sufficiency, environmentally sustainable practices and economic growth. The Commission (2000c, p. 5) has itself stated that food security and food self-sufficiency are not necessarily related:

> It must be stressed that at the national level food security is *not* synonymous with food self-sufficiency. Promotion of self-sufficiency as an aim in its own right can be counter to improving food security if it hinders policy and institutional reforms that are designed to encourage broad-based growth.

The EU is thus in a difficult position, in that any rural development programme it produces must be seen to be both effective in reducing poverty levels and achieving food security, whilst simultaneously ensuring 'environmentally-friendly' outcomes. As demonstrated in the previous chapter, the Polity maintains that boosting exports and trade liberalisation together represent the clearest strategy for combating poverty. This strategy forms the basis for its funding of the ACP agricultural sector, despite the fact that EU-commissioned sustainability studies are largely ambivalent regarding this approach (Kirkpatrick and Lee 1999, p. 117).

EU agricultural policy

The spectre of the Common Agricultural Policy (CAP) looms large over EU links with developing countries. As previously outlined, developed countries, and particularly those within the EU, heavily subsidise and protect their agricultural sectors. The CAP accounts for a greater slice of the EU budget than any other single expenditure, with EU farmers benefiting from approximately €50 billion in funding ($65 billion) in 2007 (European Commission Budget 2007). Farm subsidies account for 37 per cent of the value of the agricultural sector in the EU. Furthermore, levels of EU support are rising, with 2002 levels standing €5 billion ($6.5 billion) higher than they were five years previously in 1997, this despite a Uruguay Round commitment to reduce agricultural subsidies (Godfrey 2002, p. 6). The accession countries have seen further budgetary increases for the CAP.

For many developing countries, the CAP represents all that is unequal about North–South relations and serves to highlight the double standards being perpetuated by developed countries with regard to trade liberalisation and the dismantling of outmoded tariff regimes. The CAP is one of the few issues uniting free trade advocates and environmental and social justice campaigners. Trade-distorting subsidies, environmental degradation, overproduction and the 'dumping' of surpluses on developing countries have all served to generate reams of negative publicity for the Policy. Despite comprehensive reform proposals being advocated as far back as 1968, the CAP has endured, largely at the insistence of the French. However, the 2003 Mid-Term Review (MTR) promised a radical overhaul of the CAP so as to respond to the concerns of both environmental lobbyists and the WTO. The original proposals for the MTR put forward by the Commission were, on closer inspection, neither as dramatic nor as far-reaching as had been touted. Moreover, these proposals were subsequently watered down by the Council, again largely at the insistence of the French, with the result that few critics of the CAP were appeased by the new measures.

The origins of the CAP

The French were instrumental in formulating the CAP and have fought long and hard to ensure that it remains a cornerstone of EU domestic policy. In exchange for agreeing to open up its markets to German manufacturing, France insisted on a highly regulated agricultural regime based on protectionist pre-War French policies. The CAP, when it finally came into effect in 1962, was arrived at only after long, protracted and often acrimonious negotiations between the six original members of the EEC. According to the Treaty of Rome, the CAP was to fulfil the following remit (Howarth 2000, p. 4):

1. To increase agricultural productivity.
2. To ensure a fair standard of living for the agricultural community.
3. To stabilise markets.
4. To assure availability of supplies.
5. To ensure reasonable prices to consumers.

In the context of the time, the CAP was in line with policies being put forward in Britain and the USA. Farmer's incomes were perceived to be vulnerable if exposed to market forces and thus the CAP can be viewed as an attempt to ensure income parity between agriculture and other sectors.

Initially the CAP was highly effective and ensured European agricultural self-sufficiency. However, by the 1970s, it was responsible for the generation of vast surpluses as a result of guaranteed prices for farmers and the linking of subsidies to production levels. Images of 'wine lakes' and 'butter mountains' stand as enduring symbols of EU inefficiency and wastefulness. The effects of tying subsidies to production have also had a negative impact on the natural environment. In the UK alone, an estimated 300,000 miles of hedgerows have been lost as farmers have attempted to increase their acreage (Green and Griffith 2002, p. 16). Furthermore, the abundant use of artificial fertilisers and pesticides has created additional problems including declining wild bird populations and the pollution of watercourses.

The sheer cost of implementing the CAP has ensured a measure of controversy. The Policy currently accounts for close to half of the EU's entire budget.[3] The high cost of maintaining the regime has led to disputes between member states, the most notable of which was Margaret Thatcher's demands for a rebate for Britain in the early 1980s. These costs have been borne by EU consumers and taxpayers who, it was estimated in 2007, contribute €146 ($186) annually for every man, woman and child living within its borders (European Commission 2007b). This 'contribution' is extracted by virtue of higher food prices within the EU, together with increased taxes. The British Government has assessed the overall UK contribution at £9 billion ($17.1 billion) annually (House of Commons Hansard Debates for 8 June 2005).

The impact of the CAP

The CAP has had a profound effect on a number of developing countries. As a result of numerous CAP-engendered subsidies and tariffs, EU farmers have been able to 'sidestep' high production costs and nullify the comparative advantage in the agricultural sector held by developing countries. Studies suggest that the CAP has resulted in a domestic non-grain sector that is eight times larger than it would be if exposed to market forces (Borrell and Hubbard 2000, p. 20). The milk and grain sectors are likewise 50 per cent larger than they would be ordinarily.

As a result of the CAP, the EU is now responsible for 15 per cent of all agricultural exports (Malzbender 2003, p. 1). This statistic represents a notable feat, given that, under free market conditions, the EU would be wholly uncompetitive in the agricultural sector as a result of the high production costs prevalent within its borders. To produce one tonne of white sugar in the EU costs $660 as opposed to $280 in Malawi or Zambia. The CAP white sugar is thus approximately three times more expensive to produce than that produced in LDCs (Raworth 2002, p. 6). Yet, despite this sizeable disparity in terms of comparative advantage, the EU manages to produce an annual five tonnes of sugar surplus which is then exported (Watkins 2004, p. 1). The result has been that, despite production costs that are three times higher than those of some of its competitors, the EU is currently the world's largest exporter of white sugar (Green and Griffith 2002, p. 8). In addition, export subsidies ensure that this sugar is competitive on global markets. However, the cost of these subsidies to the EU consumer is prohibitive because, in effect, it costs the EU €3.3 ($4.3) to export €1 ($1.3) worth of sugar (Watkins 2004, p. 1). The sugar industry is further shielded by a system of high tariffs and generous production subsidies. In comparison to the average of 5 per cent for most EU tariffs, tariffs protecting sugar are effectively 140 per cent (Raworth 2002, p. 6). This pattern is replicated in other sectors. Famously, in 2002, CAFOD estimated that the average cow in the EU was subsidised at a rate of $2.2 per day – effectively more per head than that earned by nearly half of the world's population (Green and Griffith 2002, p. 7).

The CAP allocated €2.5 billion ($3.25 billion) to the dairy regime in 1999. The Organisation of Economic Cooperation and Development (OECD) estimates that if the proceeds from tariffs, higher domestic retail prices and subsidies are combined, EU countries provided an additional $17.4 billion to the dairy sector in 1999 (cited in Green and

Griffith 2002, p. 11). As a result, the EU is able to export skimmed milk powder at half of its production cost. Estimates suggest that, outside of Latin America, the CAP has decreased the production of milk products by 50 per cent.

The total annual cost of the CAP to the world economy is $75 billion, although the EU itself accounts for two-thirds of this figure (Borrell and Hubbard 2000, p. 21). South Africa loses $60 million a year in lost revenues as a result of EU sugar policies. Similarly, Mozambique and Malawi are estimated to have lost $38 million and $32 million respectively in 2004. In effect, this means that for every $3 that the EU directs towards Mozambique in aid, $1 is taken back by the sugar regime (Watkins 2004, p. 2). These sums are not large in European terms, but are significant when viewed from an LDC perspective. Furthermore, the agricultural sector in Europe employs just 4 per cent of the EU workforce; in countries such as Mozambique up to 90 per cent of the population is dependent on this sector (Lynton-Evans 1999, p. 1). Moreover, the sugar industry is the single largest employer in Mozambique, providing 23,000 jobs, yet, when CAP and EBA restrictions are factored in, it means that Mozambique is foregoing a potential $106 million per year in lost revenues (Godfrey 2002, p. 3).[4]

The CAP and the WTO

Despite the disparities engendered by the CAP it has been pressure applied by the WTO, rather than by socio-environmental campaigners, that has prompted reform of the regime. Notwithstanding the fact that the GATT nominally covered agriculture, the sector remained largely free from the provisions of the latter until the conclusion of the Uruguay Round in 1994, mainly as a result of US and EU pressure. Before 1994, developing countries played only a very minor part in dictating the direction of the multilateral trading regime. However, when negotiations for the new round began in 1986, many developing countries, and particularly those in Latin America, took a hard line on agriculture and demanded that action be taken to reduce uncompetitive practices engendered by developed countries.

Developed countries extracted a 'peace clause' during the negotiations, which was to run until 2003, and which would allow them to overhaul their agricultural sectors over an extended period. It is for this reason that the Commission initiated its Mid-Term Review (MTR) of the CAP in 2002, the details of which were finalised in June 2003. The main aim of the MTR was to ensure that the CAP adhered to WTO conventions. In order for this to occur it was necessary for the EU to make a number of changes to the CAP so as to ensure that backing for agriculture was classified by the WTO as non trade-distorting, 'green box' support rather than being assigned 'blue' or 'amber box' status. This essentially entailed the Commission finding ways of decoupling subsidies from production. The solution, according to the MTR, was a policy shift towards 'single farm payments' in order to promote 'rural development'. The Commission (2002c, p. 28) has argued that by so doing it will fulfil its WTO commitments whilst simultaneously improving environmental, health and safety and animal welfare standards:

> The approach proposed by the Commission contributes to the EU's strategic goals and commitments in the international field by making new efforts to promote sustainable development and coherence. In particular, by re-orientating support towards more extensive agricultural practices and minimally trade-distorting domestic support...The direction of the EU is clear: reductions in production-distorting support and an increased focus on food safety and quality, rural development, and environmental services for society at large.

At a press conference held on 26 June 2003, immediately after the Council had come to an agreement on the proposed reforms, Franz Fischler (2003b), the then head of the Directorate-General (DG) of Agriculture, Rural Development and Fisheries, stated

> Today marks the beginning of a new era. European agricultural policy will change fundamentally. In future, our products will be more competitive, and our agricultural policy will be greener, more trade friendly and more consumer orientated...The reform's message to the world is clear: today we have largely said goodbye to an old system of support which distorted trade. The new agricultural policy is trade friendly, particularly as regards its effects on developing countries.

However, a study of the Commission's proposals for the MTR reveals scant regard for the effects of the CAP on developing countries. In fact, as highlighted by NGOs such as the Catholic Fund for Overseas Development (CAFOD), just one sentence in the entire communication to the Council and the European Parliament relates to developing countries.

Despite the Commission's (2002, p. 28) claims that reforms would 'improve market opportunities for developing countries', closer inspection suggests that, due to the skewed nature of the changes, the effects of the reforms will be detrimental to a number of ACP countries. First, the MTR does not address market access, which is only due for consideration when the CAP comes up for full review in 2013. Furthermore, consideration of the dairy and sugar regimes was largely excluded from the discussions. Even if the remainder of agricultural support is decoupled from production, there is little evidence to suggest that this will prevent overproduction and dumping. Data from the USA suggest that decoupling does not lead to a decrease in production, and that in fact the opposite is true. Similarly, the EU cereal sector, decoupled since 1992, has continued to expand and is projected, by DG Agriculture's own forecasts, to continue in this vein (European Commission 2003c, p. 15):

'The impact of the CAP reform proposals on the income of the agricultural sector would be very modest ... some 8.5 percent higher than in 2001 (in real terms and per work unit).'

Developing countries also stress that, despite the reforms, the single farm payment will result in an average annual payment of $16,000 to every farmer in the EU (Green and Griffith 2002, p. 7). It is arguable that such a subsidy, no matter what WTO box it falls within, is bound to be trade-distorting and will continue to put ACP countries at a disadvantage.

Some ACP countries do, however, benefit from the CAP (see Chapter 5) and it is these countries that stand to lose the most from the MTR reforms. As a result of the high tariff barriers that protect European agricultural produce, domestic prices for agricultural goods within the EU are very high – sometimes up to three times higher than the world average (Stevens 2003, p. 11). This means that countries with preferential access to EU markets can profit from such high domestic prices. The MTR reforms will bring about a depression in domestic prices over time but will not affect the levels of preferential quotas and other restrictions which limit quantitative access to EU markets. Developing countries with preferential agreements with the EU (including the ACP) will thus receive less for their exports and, furthermore, will be unable to offset this loss by increasing sales.

Even if the CAP were to be abolished overnight, the potential gains would by no means be universally shared. The greatest protection against agricultural imports is levied on the so-called 'temperate zone commodities' such as wheat, coarse grains, oilseed, fruit, vegetables and livestock. Southern Africa farms a number of temperate zone products and would be likely to benefit from CAP reforms in this respect. However, the bulk of this produce is farmed in developed countries and is often not suited to production in tropical developing countries. If protection for these products were to be lifted, the main beneficiaries would likely be countries such as Brazil, Argentina, Australia and New Zealand (Schmidhuber and Raney 2003, p. 239).

EU-funded projects in ACP member states

EU support for the agricultural sector in ACP countries accounts for a relatively significant proportion of allocated funds.[5] The Commission has financed a number of country strategy evaluations in order to assess the efficacy of EU policies in this sector. For example, in Namibia, under the Seventh and Eighth European Development Funds (EDF), a total of €28.8 million ($37.4 million) was set aside for agriculture and rural development out of a combined total allocation of €189.3 million ($245.95 million) (Bird *et al.* 2001, p. 15). Similarly, Mozambique received €30 million ($38.98 million) under the Eighth EDF for rural development as well as a further €43.6 million ($56.65 million) in order to establish a Food Security Programme (Montes *et al.* 2000, p. 17). While a number of programmes are targeted at people 'on the ground', a significant percentage of agricultural sector funding has gone towards institutional capacity building. In Namibia, financing has sponsored the Support to Agricultural Marketing and Trade Negotiating (SAMTRAN) project, which has enabled the Government to develop its negotiating capacity in international forums. As a result, between August 2001 and September 2002, the Namibian agricultural trade negotiating team attended 40 international trade-related meetings. Before SAMTRAN, Namibian delegates rarely attended such events and seldom made any submissions to them. As a result of funding for training and travel expenses, Namibian delegates are now actively participating in WTO Agreement on Agriculture discussions. Funding has also been provided for Namibia to introduce a Standards Bill and an Accreditation Bill which will enable exporting farmers to meet stringent Sanitary and Phytosanitary measures relating to agricultural produce, thereby boosting their trade with developed countries (European Commission 2003b, p. 4). Similarly, in Mozambique, nearly half of the funding towards the Food Security Programme, €20.4 million ($26.52 million), took the form of budget support for a number of government ministries (Montes *et al.* 2000, p. 22).

EU-commissioned evaluations of this institutionally based strategy suggest that the impact of this funding on poverty levels in the southern African region has been mixed. In Mozambique, the impact of EU policies has been 'uncertain' (Montes *et al.* 2000), whilst in Namibia the effects of EU funding are 'likely to show up only over the longer term' (Bird *et al.* 2001). Furthermore, it appears that sustainable development is not always prioritised. The report on EU funding to Mozambique states (Montes *et al.* 2000, p. 27) 'Although there has been an increased focus on poverty both by the government and the EC, there has been limited attention given to environmental protection...There has not been a systematic review of the likely environmental impact of EC projects.'

In Namibia, the EU has chosen not to focus on environmental policy and has left this area to bilateral donors such as Norway and Denmark. While environmental impact assessments studies are integral to all major EU-funded projects, there is a sense, based on the above, that the environmental aspects of the EU's sustainable development approach are not always prioritised, despite Cotonou's (Article 1) claims that 'the principles of sustainable management of natural resources and the environment be applied at every level of the partnership'.

EU-funded food security programmes

The extent of any country's food security is generally tested when unexpected 'shocks' such as severe drought or flooding occur. Such crises undermine food production and the movement of food to markets, and have an adverse effect on availability. The EU has taken steps to address both the needs of ACP countries vulnerable to such shocks and the resulting shortages. Early Warning Systems (EWS) form a key aspect of this strategy. Analysts overseeing these EWS make year-on-year comparisons of rainfall levels, satellite imagery assessing levels of planting, the extent of harvesting, food prices and general climatic conditions. This information can then be used to predict the availability of food in a given country or region. The EU is involved in funding EWS at the national level in ACP countries such as Ethiopia and Malawi, as well as on a global level with funding to the FAO-run Global Information and Early Warning System (GIEWS). The Commission has also set aside funds for the establishment of 'village grain banks' in Sahel countries such as Burkina Faso.

The success of such programmes is vital if food security in shock-prone areas is to be assured. The 2001–03 famine in southern Africa affected millions of people and has highlighted just how vulnerable those in rural areas are to natural crises. The famine was most apparent in six SADC countries: Lesotho, Mozambique, Malawi, Swaziland, Zimbabwe and Zambia, and estimates suggest that by March 2003 approximately 14 million people had experienced severe food shortages (Southern African Development Community 2003). The SADC Secretariat (2003) laid the blame for the crisis on two years of drought, inadequate investment in the agricultural sector and the impact of HIV/AIDS. The fact is, however, that the impending crisis remained undetected for far too long despite there being a number of programmes in place to forestall such a catastrophe. NGOs had tried to highlight food shortages in a number of these countries but these warnings were generally ignored.

In July 2001, the Famine Early Warning Systems Network Malawi (FEWSNET), funded by United States Agency for International Development (USAID), reported that although maize production was down compared to previous years, the shortfall would be offset by other cereal crops and tubers. The report went so far as to suggest that an overall national food surplus of 438,000 metric tonnes (mts) would eventually be generated (cited in Sawdon 2002, p. 6). This figure was adjusted downwards to 263,000 mts a month later. The Government of Malawi, unaware of the potential scale of the problem, sold off 165,000 mts of its maize reserves before September 2001, leaving virtually no cover for contingencies. In October, the NGO Save the Children UK found that in the Mchinji District of Malawi the price of maize had risen by 340 per cent while production had fallen by 40 per cent (cited in Sawdon 2002, p. 6). Save the Children attempted to alert international donors but found that many, including the EU, believed its reports to be exaggerated. A joint FEWSNET Malawi, World Food Programme and European Commission Food Security Programme assessment in November 2001 found that, although maize production was down from previous years, food security was not threatened (cited in Sawdon 2002, p. 7). The effect of this report was to deflect attention away from Malawi's agricultural problems. By the beginning of 2002, the agricultural sector in Malawi was on the point of collapse. Maize is the staple food of the region and although other cereals and tubers are indeed cultivated, it is only on a relatively small scale. Desperate to generate income, farmers began to sell off their livestock in large quantities, severely depressing prices, while maize began to retail at levels between 400 and 500 per cent higher than normal. In February, FEWSNET Malawi, the World Food Programme and the European Commission Food Security Programme conducted an additional assessment, which found the situation was indeed severe (cited in Sawdon 2002, p. 9). The EU eventually agreed to a €310 million ($402.95 million) emergency aid package in order to mitigate the effects of the food shortages (European Union at the United Nations 2002).

It has become a major concern that, while the EU currently funds a number of food security programmes in conjunction with the Southern African Development Community (SADC) to forestall just such eventualities, in many quarters the famine came as a surprise. As stated, the EU felt that warnings from Save the Children in Malawi exaggerated the extent of the problem, given that evidence from early warning programmes in the region failed to corroborate the NGO's findings. Additional project failures in this respect include the SADC Regional Remote Sensing Unit (RRSU), established in 1988 with funding from Japan and technical assistance from the FAO in order to provide early warning indicators and to facilitate natural resource management. Since June 1998, the SADC member states and the EU have jointly funded the programme. The RRSU utilises satellite images to monitor crop production, rainfall occurrence and vegetative cover and thus should, in theory, enable analysts to make provision for natural shocks such as droughts. Despite the availability of this technology, the RRSU failed to anticipate the consequences of the drought. The EU funding also helped in the establishment of the SADC Food Security Training Programme, developed both to supply technical support to food security agencies in the region and to enable the seconding of experts to assist relevant agencies in individual member states. Clearly, on consideration of the extent of the 2001–03 famine, these programmes have proved ineffective. Furthermore, while the EU offered €310 million ($402.95 million) in aid in response to a UN-SADC appeal for assistance, funding for the proposed Agricultural Recovery Programme was slow to materialise (Southern African Development Community 2003).

The failure of the early warning programmes, as well as poor assessment of the situation by the European Commission Food Security Programme, suggests that, although funding is being channelled into vital areas, the programmes themselves have been deficient in some respects. Poverty levels are unlikely to decrease whilst hunger remains a core concern for those living in such regions of the ACP. Moreover, the net result is that many inhabitants of these areas, in the interests of survival, will be forced to further exploit the natural resources available to them, thereby exacerbating rural environmental degradation and trapping those concerned in a vicious cycle producing ever-diminishing returns. In this instance it is not that EU funding encourages environmental degradation but, rather, that EU initiatives have been unsuccessful in combating poverty and related environmental pressures.

Conclusion

If developing countries cannot benefit from comparative advantages such as low labour costs, then it is likely that EU programmes designed to facilitate an export-orientated agricultural sector in the ACP countries will have little overall impact in boosting rural economies and self-reliance. However, as alluded to in Chapter 5, the solution is not a simple one. With regards to bolstering trade and promoting self-reliance, the CAP produces both winners and losers amongst the ACP countries. A simple dismantling of the CAP might well not produce the desired results but substantial reform of the Policy, based on clear development goals and objectives, must be considered with some urgency. Regarding the CAP MTR, the fact that developing countries only merited a solitary sentence in the Commission's communication to the Council and the European Parliament suggests that ACP needs are being poorly served.

However, less contentiously, effective food security programmes are an area that could be of great benefit to the ACP, especially the African members. The 2001–03 famine in southern Africa highlighted a number of shortcomings in this regard and, if lessons can be learned, it may be possible to avert such catastrophes in the future. The basic platform for any poverty reduction strategy is the guarantee of food security and the EU needs to ensure that sponsorship of such programmes continues and, moreover, that greater funding is made available.

7
Energy Provision, Poverty Alleviation and Sustainable Development

Introduction

The Millennium Development Goals (MDGs – see Table 7.1), whilst explicitly aimed at poverty alleviation, all contain an environmental component. The eradication of hunger has obvious links to soil fertility and the need for functioning ecosystems. Less obvious are the environmental links to the MDGs of reducing child mortality, combating disease, improving maternal health and promoting universal primary education. Chapters 7 and 8 stress the importance of environmental factors in determining development outcomes, and how these best be acknowledged and addressed. The following statistics should be considered (Steele *et al.* 2002, pp. 10–14):

1. Up to 30 percent of all disease in Sub-Saharan African can be attributed to environmental factors.
2. Waterborne diseases kill nearly 3 million people a year, the majority of whom are young children.
3. Indoor air pollution and the effects of carrying heavy loads of water and firewood affect the health of women in developing countries and can lead to complications during pregnancy.
4. The time devoted to gathering water and firewood by young girls frequently results in them forgoing such educational opportunities as are available to them.

At the World Summit on Sustainable Development (WSSD), held in Johannesburg in 2002, the EU committed itself to a number of initiatives

Table 7.1 UNDP selected country energy indicators

Country	Electricity consumption per capita (kilowatt hours)
Angola	135
Bahamas	6,084
Burundi	25
Chad	12
Democratic Republic of Congo	91
Fiji	625
Mozambique	378
Niger	40
Norway	26,640
Rwanda	23
Sierra Leone	54
South Africa	4,715
Tonga	340
Trinidad and Tobago	4,422
United Kingdom	6,614
United States	13,456

Source: Watkins (2005).

aimed at targeting some of the problems highlighted above, unveiling, amongst other programmes, the EU Energy Initiative (EUEI). The Initiative identified increased energy provision as fundamental to the achievement of the MDGs. The extent of ACP 'energy poverty' is clearly outlined in Table 7.1, while the successes of EU programmes aimed at facilitating development through energy provision are duly assessed below.

Linking energy provision and poverty alleviation

Empirical evidence suggests that improved access to energy resources can result in meaningful welfare gains (Reddy 2002, p. 117). For the poor, securing adequate energy provision often depends on labour intensive exercises such as gathering firewood and making charcoal, or entails working with 'dirty' fuels such as coal and paraffin that can result in respiratory problems and even premature death amongst the very old and the very young. Furthermore, the poor pay a comparatively high cost per unit of energy. Light from a kerosene lamp can cost up to 70 times more than light generated by an electrical source. Similarly, candlelight is approximately 150 times more expensive than the cost of grid-fed electrical lighting, while battery-powered illumination is 10 to

20 times more expensive (Department for International Development 2002, p. 7). Even where fuel is free, as in the case of biomass, the associated time costs are often prohibitive. Women in parts of Sub-Saharan Africa, in order to provide for their family's energy needs, often carry loads of up to 20 kg of wood across an average distance of 5 kilometres a day (Practical Action 2005, p. 4). Studies have shown that even small gains in energy availability can help offset these costs. For example, 100 watts per capita, approximately one-tenth of the per capita energy needed to power a home in a developed country, can provide basic home electrification (Reddy 2002, p. 120). Access to fuels such as liquid petroleum gas (LPG) can cut indoor air pollution and reduce reliance on biomass for cooking and heating. These improvements can represent considerable Human Development Index (HDI) gains: increased access to energy allows for the creation of mechanised micro-industries, it facilitates after dark activities such as adult education courses, it improves access to clean, pumped water and enables rural clinics to store vaccines in refrigerated conditions. Part of achieving the MDG of halving poverty by 2015 would thus entail providing the poor with far greater access to energy resources.

Sub-Saharan Africa in particular is an energy-poor region. The majority of people rely on 'traditional' sources of energy; namely fuelwood and other forms of biomass. Access to modern forms of energy, and, specifically, electricity varies across the region, from 66 per cent in South Africa (Spalding-Fecher 2002, p. 21) to just 7.2 per cent in Mozambique (Greenpeace and ITDG 2002, p. 55). Only 1 per cent of rural households in Mozambique have access to electricity. Consequently, areas surrounding significant urban settlements have been deforested, in some cases encompassing a radius of up to 200 kilometres (Greenpeace and ITDG 2002, p. 61). As a result, given that over 80 per cent of people living in SSA are heavily dependent on biomass for energy production, energy consumption in Africa is actually declining (European Commission 2004c, p. 2; Practical Action 2005, p. 4). In 2000, SADC countries were responsible for just 1.4 per cent of global energy consumption. However, even this figure belies the extent of the region's 'energy poverty', given the fact that South Africa alone consumes 83 per cent of this total (Energy Information Administration 2006, p. 1). South Africa consumes 46 per cent of all the electricity produced on the continent (Practical Action 2005, p. 8).

Energy is also a problem for many living on island states. The costs involved in shipping fuel to remote areas means that electricity within Caribbean and Pacific states costs an average of 20–50€c/kWh as opposed to an international average of just 5€c/kWh. Such costs mean that some Caribbean ACP island states spend as much as 75 per cent of their foreign currency earnings on fuel (European Commission 2004c, p. 5). The discrepancies between energy-rich Trinidad and Tobago and the remainder of the Caribbean island states are marked. Studies suggest that the high cost of energy in Jamaica is injuring the flow of FDI into that country (Parsan 2006, p. 31). However, unlike in SSA, this problem has not resulted in across the board 'energy poverty'. Pacific island states such as the Cook Islands, Nauru, Tonga and Tuvalu have rates of electrification of between 85 and 100 per cent (Wade 2005, p. 6).

The EU has targeted energy provision as one of its development priorities and, in the build-up to the 2002 WSSD in Johannesburg, it put forward an energy programme designed to aid the alleviation of poverty and the promotion of sustainable development in all ACP states. It has exhibited a willingness to invest in ACP energy initiatives in the past, and the new energy proposal demonstrates a strengthening of this commitment. However, if climate change is to be arrested, then developing countries cannot mirror the energy production methods of countries in the North, with their reliance on fossil fuels. Rather, alternative strategies must be developed in order to mitigate the impact of increased energy consumption on the environment. A significant barrier to any movement in this direction is that the most successful electrification programme in the ACP, that of South Africa, has been achieved with scant regard to sustainable development principles.

The EU energy initiative

The initiative identifies energy as a central component of any poverty reduction strategy or sustainable development enterprise. Those supporting the initiative stress that, without adequate energy supplies, economic growth is difficult to achieve, as are effective communications networks, water supplies and heating and cooling systems. Furthermore, a lack of energy security has a number of associated health risks, including respiratory problems caused by the burning of biomass, coal or paraffin (European Commission 2003e, p. 4). It is estimated that there are approximately two billion 'energy-poor' people and, if the MDG of halving poverty by 2015 is to be achieved, then the energy sector is one that needs to be addressed. Athanassios Theodorakis (2002), the then

Deputy Director-General for Development, underlined this point in a speech at the European Union in July 2002:

> The main objective of the EU Energy Initiative to be launched at WSSD is to facilitate the achievement of the Millennium Development Goal of halving the number of people in extreme poverty by 2015, through the provision of adequate, affordable, sustainable energy services. We hope the initiative will provide a catalyst on which future action ... can be built.

These sentiments were echoed in a subsequent speech to the Global Forum for Sustainable Energy by the then Director of DG Development, Friedrich Hamburger (2002):

> Improving access to appropriate, affordable, clean energy services is a necessity if we are to achieve the Millennium Development Goals. The linkages between access to energy services and poverty eradication are well known. Without better access to energy it will be hard to develop businesses, homes, schools; health centres will remain without adequate energy for lighting, communication and cooling, streets will remain dark at night, women and children will continue to be exposed more than anyone to indoor air pollution from the use of biomass for cooking and heating. We thus believe that access to energy is an essential underpinning element in achieving poverty reduction and sustainable development.

Before the launch of the EUEI, the EU had 'earmarked' €700 million ($910.23 million) for injection into energy programmes in developing countries and wished to use the initiative to build on this figure (Danish Presidency of the EU 2002). It has therefore made the initiative available to all ACP countries. The EU has further emphasised the importance of regional groupings to the successful implementation of the initiative. The EUEI will attempt to address a wide-ranging array of issues including more efficient uses of biomass, rural electrification, decentralised energy services and sources of renewable energy. The EU has stressed that it is conscious of the links between environmental degradation and energy generation. In consequence, it has expressed itself willing to support the application of sustainable development principles to the EUEI wherever possible.

In keeping with the spirit of the Cotonou Agreement, the initiative is one based on 'partnerships'. The EU is anxious to emphasise that donor aid alone will not fund energy programmes. Instead, partnerships between the EU, the private sector and the governments concerned must be established in order to effectively promote the dissemination of sustainable energy services to those living in poverty. The EU advocates that interested parties raise additional monies in the form of grants, loans, risk guarantees and equity investments. The EUEI envisaged a two-year planning period in which to establish and cement partnerships: Phase 1, now complete, followed by a long-term, 'open-ended' programme based on the objectives established by the partners in the first phase (Phase 2). In 2004, under the auspices of the EUEI, the Commission proposed funding of €250 million ($325 million) for the formation of an ACP-EU Energy Facility (European Commission 2004c, p. 2). The Facility was approved by the joint EU-ACP Council of Ministers in June 2005 with a moderately lower budget of €220 million ($286 million) (ACP-EU Energy Facility 2006). The impact of the initiative will only truly become apparent closer towards the Millennium Goal date of 2015. However, certain conclusions can be drawn both from the nature of the programme and through analysis of the efficacy of EU programmes which have preceded the EUEI. For this reason, an overview of domestic EU energy policy and its spillover into development strategies represents a useful starting point where the assessment of the potential of future initiatives is concerned.

EU energy sector objectives 1990–2010

A study undertaken by the European Environmental Agency (2002, p. 8) highlights a number of concerns pertaining to energy consumption trends within the EU and the impact that these might have upon the environment. The study reveals that, despite progress in a number of areas, EU patterns of energy usage still represent some cause for concern. EU emissions of greenhouse gases between 1990 and 2000 showed a noticeable decline. However, the rate of this reduction has since slowed and, as a result, it appears that the EU will be unlikely to meet its Kyoto target of an 8 per cent reduction, from 1990 levels, in greenhouse emissions by 2010. The main culprit for this slowdown in emission cuts is the transport sector, which is offsetting gains made in other, non-energy-related areas. A further study has shown that, of the pre-2004 member states, only Britain, France, Germany and Sweden are on track to meet their Kyoto targets (European Environment Agency 2004b). It has been demonstrated that energy consumption has increased in all sectors barring manufacturing. Moreover, the decline in the manufacturing sector's energy consumption can be mainly attributed to the relocation of energy-intensive industries to non-EU countries and a switch to low energy-intensive industries within the EU as a whole (European Environment Agency 2004a, pp. 9–10).

Electricity production in particular represents cause for concern. One unit of electricity often necessitates the consumption of between two to three units of other energy sources such as coal. As the EU derives more than half of its electricity from the burning of fossil fuels, it is likely that growing demand for electricity will result in an increase in carbon dioxide emissions – unless there is a dramatic increase in the market share of electricity obtained from renewable sources. However, between 1990 and 1999 there was little discernible development in the renewable energy sector. The European Environment Agency (2002, p. 21) study suggests that, if the EU is to meet its renewable energy target of 12 per cent of total energy consumption by 2010, growth in this sector must be doubled.

It is therefore clear that the EU's progress in dealing with energy sector-related environmental degradation remains sluggish, and that it has some way to go where placing its own house in order is concerned. Electricity remains primarily generated by fossil fuels, greenhouse gas emissions are too high, the renewable energy sector remains poorly developed and there is little incentive for EU citizens to make more efficient use of available resources. Strategies designed to rectify some of these concerns are in place. The Commission (2001a, p. 10) has set out plans for a more 'ambitious' energy tax regime and, furthermore, has emphasised the need to phase out the subsidisation of fossil fuel 'production and consumption' by 2010. However, the current state of the EU energy sector has ramifications for developing countries which, if predications pertaining to the impact of global warming are correct, will be disproportionately affected by climate change.

Energy as a development priority

In its dealings with the ACP countries, the EU has been anxious to stress its commitment to sustainable development, but the reality is that development projects linked to energy issues often mirror problems prevalent within the EU itself. The Sixth EAP specifies that 'Development policy must continue to support the sustainable management of water, soils, and forests, access to and tenure of resources, access to sustainable energy and the interaction between health, poverty and the urban environment' (European Commission 2001c, p. 59).

However, it is only recently that a sustainable approach to energy policy has become an important aspect of EU development policy – a point acknowledged by Friedrich Hamburger (2002):

> In the past, energy was not a main priority within Community development cooperation. The activities that were undertaken focused principally on large centralised power generation, with some renewable energy actions. We now recognise that we have to adapt the way we work. Firstly to consider energy for its own sake; secondly making sure that all the stakeholders are working together in order to make best use of their particular capabilities and capacities, in order to improve access of the poor to the energy services they require.

Similarly, the language of sustainable development is now regularly employed in other Commission (2002a, p. 59) documents relating to energy provision in developing countries:

> Energy is recognised as an important crosscutting element in the elimination of poverty and the achievement of the Millennium Development Goals. The strategy for sustainable energy support emphasises the need to mainstream it into the social and economic sectors (health, education, and enterprise development), as well as the environmental and other benefits of increasing use of energy efficiency and renewable energy.

Similarly, in a study jointly funded by the EU and the UNDP entitled *Energy as a Tool for Sustainable Development*, a significant degree of emphasis is placed on the need to develop renewable energy capabilities in the ACP countries. The study (Farinelli 1999, p. 3) highlights how:'[T]raditional methods of addressing energy problems in developing countries provide inadequate solutions... Fortunately, new technological solutions are available today, which, together with new insight into the essential requirements of development, may help to find the appropriate solutions.'

However, despite the expression of such sentiments, EU funding for energy programmes and projects tends to focus on 'traditional' forms of energy production. It has been shown that renewable energy programmes, because they tend to be small-scale, struggle to attract sufficient donor interest (Farinelli 1999, p. 21). As a result, renewable energy systems, such as the Regional Solar Pumping Programme in the North African Sahel, have a tendency to be viewed as interesting side-projects rather than as primary funding targets. It is important that the EU addresses such deficiencies because the ACP energy sector, although largely undeveloped in certain regions, will not be sustainable if increased demand is to be met by means of conventional fuel production.

The South African model

South Africa represents both a triumph and a source of concern for those involved in poverty reduction and sustainable development initiatives. Since coming to power in 1994, the African National Congress (ANC) Government has worked hard to expand the number of people with access to electricity. In 1990, a mere 35 per cent of the population was connected to the grid. By 2001, this figure had risen to 66 per cent. In practical terms this represents 3.1 million new connections since 1994 with the result that approximately 70 per cent of households are now connected (Spalding-Fecher 2002, pp. 12–21). The government has committed itself to universal access by 2012 (Prasad and Visagie 2005, p. vi). However, from a sustainable development perspective, the means by which this has been accomplished represents a dilemma. South Africa relies heavily upon fossil fuels to supply its energy needs and, furthermore, subsidisation has played a significant role in the expansion of its electrification programme.

Were South Africa to be forced to liberalise its energy sector, in line with WTO and EU recommendations, then the possibility of similar electrification programmes being replicated throughout the region would be doubtful. Thomas Johansson and José Goldberg articulate this concern in a 2002 UNDP publication:

> Electricity market liberalisation and privatisation could potentially threaten widened access to electricity for the poor. Private companies often have little motivation to seek out the poor with their precarious incomes and limited capacity to pay the full cost of service. Explicit policies and regulatory instruments are needed to expand service and targeted subsidies will be needed in many instances. (Johansson and Goldberg 2002, p. 10)

In local elections in 2000, the ruling ANC Government promised free electricity and water for the poorest South African communities and this pledge is being gradually implemented through a series of pilot schemes. The Cape Town City Council has also introduced a free electricity scheme for the benefit of disadvantaged communities. The cost of subsidising such largesse is estimated to be approximately R500 million ($67.76 million) (Spalding-Fecher 2002, p. 12). Furthermore, the Government has also removed VAT on paraffin in order to reduce the cost of energy provision for those who remain without access to electricity.

An additional 3.1 million connections to the grid represent a significant achievement on the part of the South African Government, one made possible only through government intervention and the fact that the country's sole electricity supplier, Eskom, remains within the public sector. This government intervention has allowed South Africa to dramatically improve its HDI for energy provision and represents a triumph for the government's social development programme. However, the South African model flies in the face of both economic and sustainable development 'sacred cows', presenting international donors such as the EU with something of a quandary. There is little indication that the EU has been swayed by this public sector success; it has continued to press for the liberalisation of the South African energy sector.

Thus, despite its success in broadening access, the ANC government has come under considerable pressure to privatise parastatals and, consequently, 30 per cent of Eskom was sold in 2006. Furthermore, this pressure has resulted in Eskom losing its monopoly over power generation in order to affect a more liberalised sector. However, the success of the government's programme has been built on its ability to intervene in the energy market and to dictate certain outcomes. Privatisation and liberalisation of the energy sector will erode this ability. In emphasising the need for the restructuring of this sector, the EU risks undermining the gains that have been made thus far.

From a sustainable development perspective, the South African model clearly represents a poor template for the rest of the ACP. South African carbon emissions are, per capita, higher than those in many developed countries (Spalding-Fecher 2002, p. 12). This is due, in part, to a number of energy-intensive industries such as iron, steel and aluminium, as well as the ready availability of almost unlimited supplies of low-grade coal. More than 90 per cent of South Africa's carbon dioxide emissions are generated by the energy sector, much of which is driven by ageing and inefficient coal-fired power stations. However, despite poor energy efficiency and high levels of emissions, South African electricity prices are amongst the lowest in the world and provide the country with a significant comparative advantage over other developing competitors. Moreover, high levels of emissions per capita make South Africa, as a signatory to Kyoto, well placed to claim funding for Clean Development

Mechanism (CDM) projects. Under Kyoto rules, developed countries can invest in projects that serve to reduce greenhouse gases in developing countries and can then offset these gains against their own emission limitation targets. There is thus little incentive for South Africa to adopt a 'greener' approach to energy issues, as there is little to be gained in monetary terms from such a step.

Nonetheless, there are signs that the South African government is prepared to investigate more sustainable alternatives, often in conjunction with SADC partners. Sizeable reserves of natural gas have been discovered off the coasts of Angola, the Democratic Republic of Congo (DRC), Mozambique, Namibia and South Africa. Of these SADC countries, Mozambique has the largest proven reserves, with 4.5 trillion cubic feet available for exploitation (Energy Information Administration 2006, p. 5). Sasol, South Africa's synthetic fuel and chemical producer, successfully negotiated the utilisation of these Mozambican reserves in order to reduce the pollution compliance costs associated with its coal-based operations. The project involved the construction of a 536-mile pipeline between Sasol's main base in Secunda and the Pande and Temane gas fields in Mozambique, at an estimated cost of $600 million (Energy Information Administration 2007, p. 5; Spalding-Fecher 2002, p. 13). Sasol and the governments of South Africa and Mozambique jointly own the pipeline, which went 'online' in February 2004.

South Africa is also broadening its nuclear power programme, which until now has been restricted to a single plant at Koeberg, outside Cape Town. In 2000 Eskom unveiled details of the construction and testing of a 'new-generation' nuclear power plant, the Pebble Bed Modular Reactor (PBMR). Approval for a demonstration reactor was eventually secured in 2003 (Ion *et al.* 2003, p. 2). Eskom is hoping to export this technology should the tests prove satisfactory. Local as well as international investors have financed the PBMR project and, controversially, from an environmental perspective, the EU has promised technical support and fuel (Nuclear Science and Technology Research and Development 2001). This expansion of the nuclear programme will result in lower national greenhouse emissions. However, safety concerns mean that this option is unpopular amongst local environmental campaign groups, such as Earthlife Africa, which continue to campaign against any extension of South Africa's nuclear energy generating capacity.

Eskom is also involved in a number of renewable energy projects undertaken as part of its South African Bulk Renewable Energy Generation (SABRE-Gen) programme, including hydroelectricity projects in the Eastern

Cape as well as a variety of wind- and solar-powered initiatives (Martens *et al.* 2001, p. 9). However, nearly all of the projects initiated by SABRE-Gen are small-scale and essentially intended for test purposes only. There is little evidence to suggest that South Africa is actively trying to build a viable renewable energy sector. Greenhouse gas emissions can only increase under such conditions.

Energy production and consumption in the ACP

If access to sustainably derived electricity is to be expanded significantly throughout the ACP, then it is imperative that the modes of production be carefully addressed. Most African states besides South Africa rely heavily on hydroelectric power, as discussed more fully below. A number of African countries, Zambia in particular, are assessing the viability of solar-powered energy and various related pilot projects, and incentives in support of this are place. Electricity production on the continent, barring South Africa, can thus be said to be relatively 'environmentally friendly', at least in terms of emissions.

As discussed above, financial institutions are often wary of investing money in small-scale, potentially unprofitable schemes, whilst the associated start-up costs tend to make them unattractive to many potential operators. However, the EU has attempted to promote small-scale, renewable energy initiatives in rural areas, of which the 'Thousand Schools Programme' is arguably a good example, given its size and ambition. In 1995, 25,900 South African schools were identified as being in need of electrification, of which 16,400 were in rural areas and would thus be suited to off-grid electrification (Klunne *et al.* 2002, p. 84). A joint South African-Dutch initiative, under the auspices of the Reconstruction and Development Programme (RDP), committed itself to electrifying 1,340 rural schools between 1996 and 1998, using photovoltaic (PV) systems. The EU-sponsored Thousand Schools Programme was viewed as an extension of this RDP project. Under its auspices, a number of additional PV systems were installed between 1998 and 2001. The EU programme offered schools a standard system designed to power three to four plug sockets (220v) and 12 to 16 fluorescent lighting tubes. Furthermore, it also promised each school an audiovisual package consisting of a television, video recorder, digital satellite decoder and satellite dish to serve as teaching aids. The project was allocated a budget of 15 million ECU ($19.5 million) (Farinelli 1999, p. 102).

An assessment, conducted between 2000 and 2001, of the viability of both programmes, yielded disappointing conclusions. Of the 48 RDP projects sampled in the survey not one was fully operational despite having only been installed between 1996 and 1998. Out of the EU projects sampled, 57 per cent were still functional, 21 per cent partially functional while 22 per cent were no longer operational (Klunne *et al.* 2002, p. 86). The most common reasons cited for the failure of the PV systems were vandalism and theft. Schools often had to resort to employing night guards and erecting fences to prevent such occurrences, with this being an additional drain on their already limited finances. An internal mid-term review by the EU team resulted in these PV systems being installed on anti-theft mounting frames and schools being forced to increase security measures. The promised audiovisual equipment has, in many cases, not been supplied due to security concerns. User-satisfaction with the PV systems was also considered in the assessment. With regard to the RDP programme, some of the headmasters interviewed stated that they were better off without the systems. However, the response to the EU project has generally been more favourable. Most of those interviewed stated a preference for a grid connection above that of a PV system as the grid was able to supply more power, was less prone to theft and vandalism and rarely developed technical problems. Despite the initial disappointments, the programme represents a commitment by the EU to aid the process of rural electrification by sustainable means and to encourage the use of solar energy. Such encouragement is necessary if the renewable sector in the region is to be bolstered.

Other EU-sponsored PV initiatives in Africa include a large-scale rural health project in the DRC. The project was responsible for the installation of 750 lighting systems and 100 refrigeration systems capable of storing vaccines in ideal conditions. Excess electricity generated by the system was employed to charge car batteries, used by the local community to power televisions and radios, and also allowed for the creation of a community television and video service. These ventures generated income for the scheme and helped to offset the considerable installation costs involved at the outset of the project (Farinelli 1999, p. 20 Box 1).

In the Pacific Island states of the ACP, solar- and wind-powered projects are also being heavily targeted. Given that, for the majority of these nations, large-scale hydroelectric developments are a physical impossibility (due to low elevation and a lack of fast-flowing rivers), wind and PV initiatives represent the only viable sustainable forms of energy production in many regions.[1] Consequently, the bulk of sustainable energy projects in this region of the ACP are centred on these sectors, an example being EUEI-financed solar-powered projects in Niue and the Cook Islands (Zieroth 2005). Similarly, by virtue of the EUEI-funded Caribbean Wind Energy Initiative, the EU has made an effort to stimulate investment in renewable energy resources in the Caribbean.

The EU has therefore been actively involved in promoting the renewable energy sector in the ACP countries. Such funding is likely to increase when financing from the CDM becomes more readily available (there are relatively few CDM initiatives in place). However, in comparison to conventional energy projects, the funding obtainable for the renewable energy sector is limited. For instance, funding for the CDM Capacity Building amongst the Private Sector in Africa Project amounts to just €455,722 ($592,595) (European Commission 2001d, p. 3). Few projects receive more than €1 million ($1.3 million), and the majority are feasibility studies or small-scale pilot projects rather than economically viable schemes capable of producing significant quantities of 'clean' power. The fact remains that nearly all of these programmes remain interesting 'side-projects' rather than serious attempts to alter the means by which energy is produced in southern Africa. Furthermore, the majority of renewable energy programmes financed by the EU in Africa, with the exception of various CDM initiatives, are localised rather than regional in nature.

Exceptions to the above are a number of hydroelectric power projects, which can be said to have a truly regional focus. The Mozambican Cahora Bassa Scheme is one such example, and Mozambique plans to expand this capacity with the completion of the $1.3 billion Mepanda N'cua Dam, 43 miles further down the Zambezi River (Energy Information Administration 2006, p. 9). Zimbabwe has authorised plans to redevelop the Victoria Falls Scheme, whilst the Inga facility in the DRC both meet domestic needs and provides a surplus for export to neighbouring states. Angola and Namibia have agreed to a joint hydroelectric venture on the Kunene River whilst the Muguga Dam in Swaziland will, it is estimated, be able to meet nearly 50 per cent of the country's energy needs, thereby reducing its reliance on South African sources (Energy Information Administration 2006, p. 11). Malawi, too, has developed a hydroelectric capacity with the opening of the $130 million Kapichira plant in 2000. The South African parastatal Eskom is overseeing many of these projects.

Hydroelectricity and poverty alleviation

The EU, through development funds or through institutions such as the European Investment Bank (EIB), is involved in the financing of a number of regional hydroelectric programmes. Mozambique can be viewed as an ideal candidate for the development of renewable energy technology projects because no national grid is currently in existence. The Cahora Bassa scheme was constructed with *apartheid*-era South African needs in mind, and produces four times the amount of electricity annually consumed in Mozambique, with the bulk of this power being exported to South Africa as a means of generating revenue. Mozambique's scattered rural population makes grid electrification expensive and non-viable. Yet, despite the costs associated with grid expansion, the government's Poverty Reduction Strategy Paper (2001–05) outlines plans for the creation of new power stations, the development of a fully national grid, and the electrification of 60,000 homes, rather than investigating alternatives such as 'stand-alone' solar-powered systems or mini-hydroelectric schemes (Greenpeace and Intermediate Technology Development Group 2002, p. 57). Furthermore, as mentioned above, there are plans for additional hydroelectric schemes on the Zambezi River.

The EU has received considerable criticism from some sectors of the environmental lobby due to its classification of large-scale hydroelectric projects as sources of renewable energy. The major dams associated with most hydroelectric schemes have a significant environmental impact and are often the cause of numerous social problems (van Gelder *et al.* 2002, p. 9). While capable of prolific output (the Itaipu scheme on the border of Brazil and Paraguay has a capacity of 12,600 megawatts) the associated costs can often be prohibitive. The Itaipu Dam cost $20 billion to complete whilst projected costs for the Three Gorges Scheme, currently nearing completion in China, are said to be as high as $25 billion (Holdren and Smith 2000, p. 77; International Rivers Network 2003, p. 11). Worldwide, more than half a million square kilometres of land have been flooded by dams and this flooding has led to the displacement of up to 60 million people during the past century. The Three Gorges Scheme alone will displace more than a million people. Dams can also have a negative effect on the health of nearby populations, often enabling the spread of diseases such as malaria and schistosmiasis.

The environmental impact on dam-building is enduring. Globally, an additional 200 cubic kilometres of water are lost each year as a result of the higher levels of evaporation associated with major dams. Moreover, dams affect the fertility of lower-lying areas by preventing waterborne sediments and nutrients from reaching these zones. Before the construction of the Aswan Dam, the Nile annually deposited approximately 10 million tonnes of sediment on the floodplain and delta. The dam, however, now traps 98 per cent of this sediment and lower-lying areas have experienced a resultant decrease in soil productivity (Holdren and Smith 2000, p. 79). Such sediment loss can also result in coastal erosion and the destruction of delta ecosystems. A study by the World Commission on Dams published in 2000 (p. 75) has demonstrated that, under certain conditions, major hydroelectric dams produce as much greenhouse gas as coal-fired power stations of equivalent output. Methane, generated by rotting vegetation, can be released in large quantities from such dams. This rotting vegetation includes submerged plant matter as well as that which is washed downstream. Climatic conditions play an important role in determining the levels of methane released. The study shows that hydroelectric schemes in northern Europe produce only one-tenth of the greenhouse gases emitted by equivalent coal-fired stations, while in Brazil such projects can actually produce higher emissions than their fossil fuel counterparts.

While climatic conditions in many parts of Africa are unlikely to cause dams to produce the high emissions of those in tropical Brazil, the perception that hydropower is a 'clean' energy source is coming under increased scrutiny. Consequently, most environmental organisations do not classify traditional hydroelectric schemes as sources of renewable energy (Greenpeace and Intermediate Technology Development Group 2002, p. 56). Despite such concerns, the EU has been a major donor behind a number of hydroelectric schemes in Africa. These include the Muela Hydropower Project, part of the larger Lesotho Highlands Water Project (LHWP), which received ECU 85 million ($110.5 million) in funding from the EU in the 1990s, and the rehabilitated hydropower schemes in Zambia, which collectively obtained loans from the EIB to the value of €32 million ($41.6 million). This pattern has been replicated in other parts of Africa, with large hydroelectric projects being financed by the EIB in Mali, Ethiopia and Egypt (Zuidberg 1995, pp. 44–47).

The LHWP highlights a number of contentious issues relating to poverty reduction, environmental degradation and the poor targeting of funds. The project, due for completion in 2020, involves the construction of five dams, 200 kilometres of tunnels through the Maluti

Mountains and two pumping stations, at an estimated cost of $8 billion (Byers 2002, p. 1). The dams will submerge 4,002 hectares of arable land and 18,709 hectares of grazing and will impact on the lives of over 20,000 people (Jayaseela 1996, p. 302). This loss of land must be contextualised: only 9 per cent of mountainous Lesotho is arable (ACP-EU Courier 2002, p. 70). Owing to such a shortage of arable land and grazing, it appears unlikely that those affected will be successfully relocated. Furthermore, the costs associated with the dam are arguably at odds with the needs of the country. There is little to indicate that the LHWP will reduce poverty levels. The hydroelectric capacity of the scheme, when complete, will be 180 megawatts, yet the Lesotho market is too small and too poor to make domestic consumption financially viable. Much of this energy will be exported to South Africa and, whilst this will bolster revenues, it is South Africa, rather than its neighbour, that stands to gain the most from the project.

Conclusion

The EU has stated that energy provision now forms part of its poverty reduction strategy in the ACP countries. The EUEI, the availability of project funding and the CDM all provide means for the expansion of energy services to those living in poverty. The EU has stressed that projects should work to alleviate poverty and that they should do so in a sustainable manner. Whether or not significant hydroelectric schemes should be classified as a renewable energy sources, the EU has invested heavily in this sector. The Polity funds a number of regional and national renewable energy initiatives, but the majority of these are small pilot projects involving sums of less than €1 million ($1.3 million). The most extensive funding is channelled towards hydroelectric schemes. There is little evidence to suggest that these large-scale projects help to alleviate poverty, and it is arguable that they engender negative social and environmental consequences for those living in close proximity to them. South Africa, meanwhile, has been successful in extending energy provision to many poor households, albeit in a manner that is neither economically nor environmentally sustainable. EU development policy requires improved targeting that simultaneously avoids the negative environmental consequences associated with the South African model. If the EU is serious about developing a renewable energy sector amongst its ACP partners, then it is essential that funding be accordingly prioritised. The EUEI is a relatively new initiative and is yet to deliver concrete results. However, its focus on delivering energy services to the energy-poor in the ACP is promising, as is the establishment of a secretariat to oversee operations. Together, these factors suggest a positive commitment that could provide the necessary dividends.

8
Access to Water and Effective Sanitation – Provision and Security of Supply

Introduction

While energy provision (Chapter 7) is crucial in terms of raising living standards, improving access to education and affording new business ventures, such opportunities cannot be fully exploited when people are continuously exposed to environment-based threats to their general health and well-being. Poor sanitation and a lack of access to clean, safe drinking water can result in exposure to cholera, diarrhoea and malaria, all of which result in millions of preventable deaths each year, the main victims being children under the age of five (Steele *et al.* 2002, p. 11). Worldwide, estimates suggest that up to 1 billion people are exposed to unsafe drinking water on a continual basis. This represents 18 per cent of the global population. Similarly, approximately 2.4 billion people are without adequate forms of sanitation (African Ministerial Conference on Water 2003, p. 4). Estimates suggest that preventable incidents of disease could be cut by as much as 75 per cent through the provision of safe drinking water and improved sanitation (European Commission 2004a). As can be seen in Table 8.1, access to water and sanitation services (WSS) varies dramatically, and it is rural dwellers who are the least likely to be able to access these services. Furthermore, global estimates suggest that severe water shortages will be a hallmark of the next two decades, with availability being expected to drop by as much as a third (UNESCO 2004). Security of supply is thus equally as important as cleanliness and safety.

The socio-economic effects of declining access to water supplies are clearly illustrated by the effects of the 1991–92 droughts in southern

Table 8.1 UNDP selected country water and sanitation services indicators

Country	**Population with access to safe drinking water (%)**	**Population with access to effective sanitation (%)**
Angola	50	30
Bahamas	97	100
Burundi	79	36
Chad	34	8
Democratic Republic of Congo	46	29
Fiji	n/a	98
Mozambique	42	27
Niger	46	12
Norway	100	100
Rwanda	73	41
Sierra Leone	57	39
South Africa	87	67
Tonga	100	97
Trinidad and Tobago	100	91
United States	100	100

Source: Watkins (2005).

Africa. In Zimbabwe, GDP declined by 11 per cent, power generation slumped by 15 per cent due to the low levels of the Kariba Dam, and food aid was necessary in order to avert a famine. Similarly, agricultural production in South Africa fell by 27 per cent, which in turn led to the loss of 49,000 jobs (Turton 2000, pp. 125–6). The link between water and development has been acknowledged by certain ACP regional blocs such as SADC, which established a Water Sector Coordination Unit in 1998 on the basis that without 'sustainable water resources management, it would be difficult, if not impossible, to attain optimal regional economic and social development' (SADC Water Sector Coordination Unit 1998, p. iv).

Prioritising water and sanitation services

The 2002 World Summit on Sustainable Development (WSSD), whilst viewed as a disappointment in some quarters as a result of what was considered to be its failure to sufficiently promote environmental protection, conveyed a commitment to the Millennium Development Goals and in so doing forced the issue of access to WSS to the fore. The EU was determined to establish its green/development credentials at the conference and was at the fore in stressing the need for a coordinated response to WSS issues. Anders Fogh Rasmussen, EU President at the time of the WSSD, emphasised the EU's commitment to providing improved water and sanitation services to the very poor:

> Water and sanitation are key to sustainable development, health, regional stability and economic stability. I believe we have the moral obligation to do the right thing: to provide clean drinking water and sanitation for every village, town and city on the planet. By doing so, every year we can save many millions of lives and prevent millions from suffering from serious diseases. (European Union Water Initiative 2002)

These views were echoed by the then Commission President, Romano Prodi:

> The global water crisis is a major threat for our planet and the future of our children. Together with our partners we are fully committed to achieving the WSSD's now-agreed targets to halve the proportion of people without access to safe drinking water and improved sanitation by 2015 ... We are committed to giving our full political support to all those working hard every day to achieve water security, in quantity and quality, for everyone on earth, today and for future generations.

A defining feature of the WSSD was the creation of approximately 300 partnership agreements between governments and stakeholders with an eye to implementing the commitments arrived at in Johannesburg. A number of high-profile WSS partnership initiatives were launched on the back of the WSSD, including the USA's Water for the Poor programme, the USA and Japan's Clean Water for People Initiative and the EU's own Water Initiative (EUWI). A number of goals were identified by the EUWI as being of critical significance, with the result that the Commission (2003a, p. 2) now wishes to appear as being on track to

1. Reinforce political commitment to action and raise the profile of water and sanitation with a view to reducing poverty;
2. Promote better water governance arrangements, including stronger partnerships between public and private sectors and local stakeholders, and build institutional capacity;
3. Improve coordination;
4. Develop regional cooperation in the application of integrated water resource management in trans-boundary waters;

5. Develop additional and innovative funding mechanisms and stimulate additional funding.

The EUWI is a partnership agreement and, in seeking to involve multiple-stakeholders, is modelled along much the same lines as the European Union Energy Initiative (see Chapter 7). The former is also designed to attempt to promote regional cooperation in this area through the encouragement of 'ownership' of projects at both national and regional levels. To this end, the need for coordination between various sector programmes is stressed. The EU has acknowledged that the EUWI, if it is to be successful, must be adequately financed and has thus allocated €1 billion ($1.3 billion) from the Ninth European Development Fund for the creation of a 'Water Facility' for the ACP countries (European Commission 2003a, p. 6).

It has been estimated that it will cost between $15–30 billion per annum in additional funding to meet the Millennium Development Goals relating to water and sanitation (Smets 2003, p. 7). The new partnership agreements, however, offer little in the way of new funding and much of what has been highlighted here is finance that was already earmarked for development. In the case of the EUWI, the €1.4 billion ($1.82 billion) on offer will be drawn from established EDF funds – it is not 'new' money. Rather, other development projects will suffer in order to finance the initiative. Similar issues become apparent when the programmes outlined by the USA are analysed. Furthermore, none of the major initiatives mentioned above offer clear operational targets and neither do they outline clearly defined goals and strategies. As a result, the success of these initiatives will be difficult to quantify, as they offer no set aims by which they can be judged.

Private sector provision

However, it has been the role of the private sector in these initiatives that has engendered the most controversy. For certain NGOs, the EUWI appears to be directed more at securing access to new markets for European water corporations than aimed at the alleviation of poverty. Critics have stressed that the EU's unwavering support for public–private partnerships (PPP) is based on a case that is yet to be made (EUWI 2003, p. 1). Privatisation of public utilities such as water provision has a number of knock-on effects. Companies are obliged to their shareholders, not their customers; the severance of state ties means that transparency may be less of a prerequisite. Empirical evidence suggests that privatisation is extremely unpopular in the South and has lead to mass demonstrations in a number of instances. More importantly, however, research has shown that privatisation is generally accompanied by price rises. Moreover, contracts tend to be long term, in the region of 25–30 years, with the result that action against failing companies is often difficult (Magdahl 2006, p. 39). Analysis of EU aid for water projects demonstrates that the majority of the funding provided was spent on initiatives that facilitated the privatisation of the sector rather than on the actual extension of water provision.

The EU has made no secret of its ongoing commitment to the involvement of the private sector and has steadfastly maintained that this approach represents the best way forward:

> The EU regards the roles of the WTO/GATS [General Agreement on Trade in Services] discussion and the water services as integral to the facilitation of greater investment in water supply and water treatment infrastructures. It is also necessary to ensure that water supply and sanitation are provided to the urban poor and to rural communities. However, it is recognised that the greater involvement of the private sector and their access to markets in developing countries gives rise to legitimate concerns about governance and regulation. The intention is not to open up markets simply to provide commercial benefits for EU companies. We are confident that EU companies can deliver high-quality services at competitive prices and with full awareness of their wider social responsibilities. (European Union at the United Nations 2003)

It has been pointed out by critics that a number of the water TNCs, including Suez, Thames Water and Vivendi Universal, were heavily influential in the shaping of the EUWI. Furthermore, the Suez Director-General for International Affairs, Ives Thibault de Silguy, a former European Commissioner, was in close contact with the architects of the EUWI and held meetings with the then Environment Commissioner, Margot Wallström, as well as the then Commission President, Romano Prodi (Corporate Europe Observatory 2003, p. 6). The size and influence of Suez, Vivendi and the like has caused concern in a number of circles. A further water company, the RWE Group, is listed along with Vivendi Universal and Suez amongst the top 300 companies on the Global Fortune 500 rankings for 2006. Suez has operations in 130 countries and, together with Vivendi, controls approximately 70 per cent of the existing global water services sector (Blue Planet Project 2002).

Suez does not have a positive track record where the delivering of water services to the poor is concerned. The company was swift to establish itself in South Africa following the end of *apartheid*, showing itself eager to take advantage of the newly elected ANC Government's pledge to privatise utilities. Nelson Mandela outlined the ruling party's new perspective in this respect, stating that 'privatisation is the fundamental policy of our government. Call me a Thatcherite if you will' (cited in Pauw 2004, p. 3). Under significant pressure from the World Bank, the South African government implemented a number of privatisation schemes in the water sector. Suez was an early multinational investor and, through a wholly owned subsidiary company, Water and Sanitation Services South Africa (WSSA), negotiated management contracts to supply water and sanitation services to the Eastern Cape Province towns of Queenstown, Stutterheim and Fort Beaufort. Between 1994 and 1999, water tariffs in these areas rose by 300 per cent and by 1996 the average household was devoting up to 30 per cent of its income to services. When customers defaulted on their payments, debt-collecting agencies were sent into the townships in order to collect arrears (Pauw 2004, pp. 3–4). Such heavy handedness led to confrontations between residents and police. A High Court order was eventually sought by the Fort Beaufort Municipality in order to release it from its contract with WSSD. Suez operations in other parts of the world, some of which were initially heralded by the company as the way forward in the provision of clean water for all, have resulted in public relations disasters such as the Buenos Aires concession, which culminated in accusations of fraud, duplicity and exploitation being levelled at the company (Santoro 2004, pp. 1–3), which nonetheless continues to list its Buenos Aires operations as a testament to the virtues of PPP (Suez 2007).

What is of concern to a number of observers is that the influence exerted by the water companies over the EU in terms of the formulation of both the EUWI and the Water Initiative is that evidence suggests, and several track records confirm, that PPP can have a negative impact on poverty levels when addressing the provision of water and sanitation services to the poor. In this respect, water privatisation in South Africa provides a useful template. Since privatisation measures were initially implemented in 1994, as many as ten million South Africans have found themselves cut-off from safe water supplies at some point. A further two million have been evicted from their homes for failing to settle their service accounts. However, most tellingly, South Africa experienced an unprecedented cholera outbreak in 2000 after communities in KwaZulu-Natal, unable to pay for utilities, turned to polluted river water to supply their domestic needs. The last recorded cholera outbreak in KwaZulu-Natal was in 1982 when 12,000 people were infected and 24 people died. The *apartheid* Government responded by providing communal taps to affected areas, which dispensed clean, free water. The majority of these taps have since been fitted with meters and can only be activated on the insertion of prepaid cards, which cost $4.02 each (Pauw 2004, p. 2). With increasing numbers of people being forced to access unsafe water, when cholera broke out in the province in August 2000, it quickly spread to the Eastern Cape and later Johannesburg, resulting in an official figure of 120,000 infections and 265 deaths. The South African Human Sciences Research Council has estimated the figure to be higher still, with 300,000 infections and approximately 300 deaths (Pauw 2004, pp. 3–4). The Government was forced to allocate $2.5 million in emergency funds to KwaZulu-Natal to fight the outbreak. Moreover, estimates suggest that diarrhoea-related illnesses in general cost South Africa $0.5 billion per year in medical expenses and a further $3.6 billion per year in lost productivity (McDonald 2002, p. 19).

The evidence from the rest of SSA does little to bolster confidence in the PPP strategy. Crucially, evidence suggests that the privatisation of water services inevitably leads to price increases. In Guinea, the cost of water to the consumer rose from $0.12 per cubic metre before privatisation in 1989 to $0.83 per cubic meter by 1996 (Bayliss 2002, p. 8). The result has been that water companies, operating along similar lines in Guinea, Gabon and Cote d'Ivoire, have made significant returns on their investments. In 2001, French water TNC Saur made profits of nearly $10 million in Gabon through its subsidiary company SEEG (Bayliss 2002, p. 11). What has angered several critics is that, since the impetus for privatisation in the water sector is essentially donor-driven, water TNCs are often able to take advantage of international pressure in order to secure contracts. This was the case in Kenya where a Vivendi subsidiary, Sereca Space, was awarded a lucrative contract to facilitate the supply of WSS. Despite not being obliged to invest any funds in infrastructure over a ten-year period, the company stood to make $169 million during the course of its contract. The company was also able to negotiate for the reimbursement of all computer and hardware costs. A sustained campaign by civil society groups meant that the government was forced to suspend the contract (Bayliss 2002, p. 6).

The case for PPP can also be said to be undermined by the fact that both Botswana and Namibia have proven to be adept at providing services through the public sector. Botswana's Water Utility Corporation (WUC) has increased access from 30,000 in 1970 to 330,000 in 1998, while the Namibian Water Corporation (NamWater), a fully owned government enterprise established in 1998, was profitable within two years and publishes all accounts to ensure transparency. Certain parastatals, such as the state water utility in Gabon, which have since been privatised, have been widely acknowledged to be effectively managed, successful operations (Bayliss 2002, p. 14). Where recent public enterprises have failed, this has occurred largely as a result of a lack of government commitment. In September 1999, an attempt at a public–public partnership (PUP) was attempted in the Odi District of South Africa's North West Province. Under the scheme, largely facilitated by the South African Municipal Worker's Union (SAMWU), the public sector was to work towards 'building the capacity of disadvantaged municipalities [in order to enable them to] deliver good quality, affordable services to the people instead of throwing in the towel to a multinational company' (cited in Pape 2001, p. 1). The plan was to bring together the Rand Water Board, a parastatal, and the municipalities of Winterveld and Mabopane, as well as certain areas under the control of the Eastern District Councils. What is more, funding, of approximately R42 million ($5.7 million), was promised by the Department of Water Affairs and Forestry (DWAF). However, the municipalities have since failed to fulfil their financial obligations and the DWAF has proved to be somewhat reticent in providing the promised funds. As a result, within a year of its establishment, the project was on the verge of collapse. The government's commitment to its preferred option, privatisation, has led supporters of the PUP in Odi to argue that the government deliberately allowed the initiative to fail in order to stress the case for PPP (Pape 2001, p. 2).

This is not to suggest that, where they have not been sabotaged by the proponents of privatisation, public utilities in Africa have all been successful in providing WSS to the poor. On the whole they have failed in this regard. The majority have suffered from bad financial management, low funding levels, low prioritisation and a lack of sufficiently trained and experienced staff, as well as a high degree of political interference, little independent regulation and corruption. What is uncertain, however, is how the private sector will be able to improve the ability of the poor to access WSS. In many instances, the poor are not prioritised because most PPP projects focus on infrastructure rehabilitation, which aids those already connected to the mains but does little to broaden access. In many LDCs, where 'markets' for water are extremely limited, good governance can be said to be more important than the quest for sector liberalisation. The strategy underpinning the EUWI is thus not one that has generated proven results and there is little evidence to suggest that privatisation will result in net welfare gains. The impressive electrification programme put in place by the South African government was made possible only because Eskom had remained within the public sector. It is thus important to assess the programmes currently in place under the auspices of the EUWI in order to ascertain whether PPP can be successful in this regard despite the examples to the contrary noted above.

The cost of public private partnerships

The Water Supply Programme Regional Centres project is a joint venture between the Commission and the German Government and involves the allocation of €44 million ($57.21 million) in funding to the Tanzanian water sector (European Union Delegation in Tanzania 2003). The programme aims to supply water and sanitation services to approximately one million people living in the towns of Iringa, Mbeya and Mwanza. The funds are to target the rehabilitation of the existing infrastructure, as well as the development and expansion of new facilities. Funding will also facilitate the installation of meters so as to ensure the 'improvement of collection efficiency' (European Union Delegation in Tanzania 2003). Sanitation is a key target for the programme; sewage from Mwanza village has been flowing, largely untreated, into Lake Victoria for a number of years. The programme has funded the installation of three new wastewater pumps and 4 kilometres of new mains, as well as the construction or rehabilitation of ten wastewater stabilisation ponds.

Central to the programme is the involvement of the private sector, which, it is argued, is vital if the project is to be sustainable over the long-term. William Hanna, head of the EU delegation in Tanzania, has stressed that the Mwanza project was one to be built on lessons from past experiences: 'we learnt from earlier projects that success depends on decentralisation and a comprehensive policy reform.... The new policy is based on decentralisation, participation of beneficiaries and the private sector' (cited in *The Business Times* 17 June 2004). However, the success of the private sector in operating Tanzania's flagship water privatisation programme has been questioned. The Dar es Salaam Water and Sewage Authority (DAWASA) has been part-privatised, with the private operating company, City Water, being responsible for billing, tariff collection, day-to-day operations and general maintenance. Details of the contract between DAWASA and its private sector partner have not been made public, which has aroused a certain degree of suspicion.

Furthermore, the privatisation of state holdings was a precondition for debt relief as well as World Bank funding. A 30 per cent tariff increase was implemented during the first year of the partnership which, together with an improved metering and collection capability, has meant an increased burden on the poor in the area (de Waal 2003, p. 2). With the poorest sectors of the population already spending as much as 10 per cent of their income on water supplies, any increases are likely to impact on people's ability to pay for these services. In order to offset this burden, a two-tiered billing system was introduced to ensure that those who consume less than 5 cubic meters of water pay less than larger consumers. However, as only 98,000 people out of a total city population of 2.5 million are DAWASA customers, such relief measures will mean little on the ground, given that the majority of people in the city are supplied by water vendors (United Nations Office for the Coordination of Humanitarian Affairs 2003). With so little financial incentive, it is difficult to see how such private sector investment will succeed in targeting the poor.

Shared water courses, conflict and food security

Water management has a clear regional dimension, the impact of which can be equally as important as the availability of WSS in alleviating the plight of those living in extreme poverty. Consequently, the EUWI stresses the need for multilateral cooperation. The European experience suggests that cooperation over the management of shared water sources promotes economic development as well as regional integration. Such cooperation also aids the easing of tensions between countries with conflicting needs and requirements where dams, irrigation and pollution are concerned. Africa has 60 trans-boundary rivers and there have been a number of potential flashpoints relating to shared watercourses, some of which continue to simmer. There have been demands for a renegotiation of the treaty that governs the flow of the Nile, because upstream countries have begun to insist on an increased input in determining the flow of the river. The Nile flows through ten African countries, all of which are bound by the 75-year-old Nile Basin Treaty drawn up by the British in 1929. The Nile Basin Initiative (NBI) was established in 1999 to try to overcome some of the obstacles presented by the Treaty, which prohibits signatories from interfering with the water flow to Sudan and Egypt. Kenya's refusal to seek Egypt's permission for planned hydroelectric and irrigation schemes on Lake Victoria in 2003 was viewed by the Egyptian

Government as a declaration of 'war' and significantly heightened tensions in the region (Pottinger 2004, p. 2).

Similarly, shared watercourses have prompted a certain degree of tension within the SADC region, most notably between Botswana and Namibia. The two countries, after repeated threats of military action, eventually settled a dispute over ownership of an island in the Chobe River by appealing to the International Court of Justice in The Hague in 1996. The Court eventually ruled in favour of the Botswanan claim (International Court of Justice 1999). However, it has been the threat from the Namibian Government to pipe large quantities of water away from the Okavango River System to its capital city of Windhoek which has sparked the most controversy. Namibia has proposed to divert approximately 17 million cubic metres (million m^3) of water via a 260-kilometre long pipeline in order to relieve the pressure on its limited water resources further south. The Botswanan Government, together with nature conservation bodies, has expressed fears that a notable reduction in water levels could affect the country's greatest tourist attraction, the ecologically sensitive Okavango Swamps (*Mail and Guardian* 17 May 2004). Despite the fact that studies suggest that such an intervention is unlikely to impact significantly on the Delta (Ashton 2000, p. 95), Tawana Moreni, the local paramount chief, allegedly advised the government to bomb the pipeline (Le Roux 1999). The situation is not helped by the fact that international water law relies on moral authority rather than any threat of enforcement. Despite the fact that it is generally acknowledged by hydro-political analysts that a 'water war' in southern Africa is unlikely, Table 8.2 illustrates the potential for possible tensions arising in the region given the number of riparian countries.

Clearly, a regional approach to dealing with the issue of shared watercourses is necessary. The SADC Water Protocol is a step in this direction and is actively encouraged by the EU. Conflict over water can only serve to destabilise the region and add to poverty levels. Moreover, water management is necessary to ensure that marginalised rural farmers are not further disadvantaged by water shortages in a region already prone to severe droughts. The annual regional average of 152 cubic metres of water per person is belied by considerable variations within the SADC grouping. In both South Africa and Mauritius, the average personal consumption is 420 cubic metres per person, whilst in Malawi this figure falls to just 20 cubic metres per capita (African Water Issues Research Unit 2000, p. 1). A 'redistribution' of some kind, on a regional basis, would help to remedy this imbalance, as would the sharing of

Table 8.2 Southern African river basins: potential for conflict

River	States	Total Area (Km^2)	Potential conflict issues
Cunene	Angola Namibia	117,000	While local water demands are low given that the river lies in a sparsely populated area, opposition to the hydroelectric project within the Epupa scheme has been voiced.
Limpopo	Botswana Mozambique South Africa Zimbabwe	423,000	South Africa, Botswana and Zimbabwe all have plans to extract water from its tributaries leaving Mozambique as the main potential loser.
Okavango	Angola Botswana Namibia Zimbabwe	586,000	Zimbabwe is not a member of the Okavango River Basin Commission (OKACOM), which represents a possible source of tension.
Orange	Botswana Lesotho Namibia South Africa	973, 000	The Orange River Basin encompasses all four countries. However, the Lesotho Highland Water Project involves only South Africa and Namibia.
Rovuma	Mozambique Tanzania	155,000	As Mozambique is denied access to much of the water of the Save and Limpopo Rivers, the Rovuma could represent an important resource for development in this country.
Save	Mozambique Zimbabwe	104,000	The major part of this river lies in Zimbabwe. Once again, Mozambique represents the main potential loser.
Zambezi	Angola Botswana Malawi Mozambique Namibia Tanzania Zambia Zimbabwe	1,234,000	Runs through eight member states of SADC.

Source: Van Wyk (1998).

expertise. Such steps are important as it can be argued that it is only by guaranteeing access to water that agriculturally dependent economies can develop and diversify. Access to water can thus be said to be on a par with access to land and, given that these issues are inextricably linked, a lack of adequate access to water could consequently imperil any development in the region.

In any shared water system, particularly one such as the Zambezi, which flows through eight countries, activities upstream have the ability to significantly impact on regions further downstream. Actions by upstream parties can affect natural processes governing such areas as vegetation, soil quality and even climate, with related impacts on grazing, irrigation and fishing. Any regulation of the flow affects levels of sediment and nutrients to downstream areas, which can in turn impact on soil fertility and food security. Mozambique is a good example of a downstream country, with nine international river basins terminating on its coastline. The country is thus at the mercy of its upstream neighbours where the implementation of best practice is concerned. Energy production, irrigation and industrial development are thus, to a certain extent, affected by policies outside of the Mozambican Government's control. The Komati River, which passes through Swaziland and South Africa before reaching Mozambique, is denuded to such an extent by South African industrial and agricultural activities (up to 50 per cent of the water in the system is removed) that salt intrusion at the river mouth now reaches between 70 and 80 kilometres inland during the dry season (Leestemaker 2000, p. 141).

In 1995, in order to overcome problems associated with shared watercourses, SADC produced both a Water Protocol and a Protocol on Shared Watercourse Systems, with the aim of governing relations between competing interests in this sector. The Water Protocol was subsequently amended in March 2000, with key elements of the Protocol including the following points:

1. Member States shall utilise a shared watercourse system in an equitable manner. In particular, a shared watercourse system shall be used and developed by Member States with a view to attaining optimum utilisation thereof and obtaining benefits there from consistent with adequate protection of the watercourse system.
2. Member states within a shared watercourse system undertake to pursue and establish close cooperation with regard to the study and execution of all projects likely to have an effect on the regime of the watercourse system.
3. Member States shall maintain and protect shared watercourse systems and related installations, facilities and other works in order to prevent pollution or environmental degradation (Southern African Development Community 1995, p. 6).

The main EU-funded project linked to the SADC Water Protocol is the Hydrological Cycle Observing System (HYCOS) initiative, which is one of a number of such projects around the world under the controlling auspices of the World Meteorological Organization (WMO). The World Hydrological Cycle Observing System (WHYCOS) was initiated by the WMO and the World Bank in 1993 in an effort to bolster coordination between countries sharing water sources. There are approximately 300 river basins and other water sources worldwide which are shared by two or more states and it is for this reason that WHYCOS was provided with a regional focus. Funding for the programme is principally provided by the WMO, the World Bank, the EU and the French Government. WHYCOS is based on the gathering and dissemination of information. Observation stations transmit data via satellite to specialist centres for analysis and interpretation, databases are regularly updated and forecasts are made based on the information available.

The SADC HYCOS project was launched in 1998 with the aim of promoting national and regional capacity in the assessment, monitoring and management of water affairs in the region (World Meteorological Organisation 2002, pp. 3–4). The programme also aimed to help effect the modernisation and rationalisation of the sector in order to ensure increased efficiency, cost effectiveness and sustainability. Key to the success of the HYCOS project is the collection and dissemination of information and for this reason a network of approximately 50 Data Collection Platforms (DCPs) was set up across the region to record flows and water quality. Such information allows states to monitor the actions of other member countries and facilitates the effective implementation of the Water Protocol and the Protocol on Shared Water Courses. The HYCOS initiative also allows members to predict and monitor the effects of drought and flooding. Given the severe flooding in Mozambique between 2000 and 2001, and again in 2007, the merits of such a programme are clear.

Conclusion

The success of EU-funded projects in reducing the number of people living in poverty and at the mercy of poor water and sanitation services must be judged against the potential of these programmes to make a significant impact on the lives of the people living in the region. EU aid to individual ACP states comes with conditionality and the inevitability of private sector involvement. It has been argued that there is little evidence to suggest that the private sector will be effective in terms of broadening access to WSS services. In reality, the evidence points to the fact that the privatisation of water utilities often results in an increased price burden that is shouldered to a disproportionate extent by the very poor. Governance, rather than privatisation, can be said to be the key factor in the deliverance of these services. Furthermore, the influence exerted by major European water TNCs in the development of the EUWI has led some NGOs to question the polity's motivation in pressing the initiative. The record of companies such as Suez in South Africa has done little to assuage peoples' fears. However, on a positive note, the SADC-HYCOS programme could substantially reduce the effects of droughts and flooding in the region, thereby improving food security and lessening the impact of such crises.

9
EU-ACP Economic Partnership Agreements

Introduction

This chapter outlines the scope of EU-ACP relations post-2000. It identifies potential areas of conflict and disagreement between the two blocs and considers the possible scenarios for both sides if ongoing trade negotiations fail to reach consensus. Attention is focused on one of Cotonou's principal features: the proposed establishment of Economic Partnership Agreements (EPAs) between the EU and the ACP countries.

According to the EU, the EPAs are 'development vehicles' (Karingi *et al.* 2005, p. 2), intended to realise the objectives of Cotonou's Article 35: 'economic and trade cooperation shall be built on regional integration initiatives of ACP states, bearing in mind that regional integration is a key instrument for the integration of ACP countries into the world economy'. WTO-compatible, they centre on the creation of free trade agreements between the Polity and regionally grouped members of the ACP bloc.

The ACP response to EPAs has been unenthusiastic – particularly in the wake of the generally articulated reticence to abandon the Lomé regime – despite a lack of any evidence that ACP countries actually developed their economies as a result of their special access to EU markets.[1] Underlying ACP concerns are both the question of reciprocity, and an underlying suspicion that the proposed agreements are more tightly focused on the EU's desire to rationalise trends in global trade than any real commitment to sustainable development or poverty alleviation. The resolution of ACP concerns is highly desirable, since the end-date for EPA negotiations, scheduled to coincide with the expiration of the WTO waiver regarding continued Lomé preferences, is December 2007.

The Cotonou template: regionalism and free trade agreements

The EU is currently negotiating EPAs with six regional ACP configurations, four African, one Caribbean and one covering the Pacific Island states, together comprising 77 countries. The groupings are as outlined in Table 9.1.

The first priority for the architects of the EPAs was to create a regime that conformed to Articles XXIV of the GATT and V of the GATS (General Agreement on Trade in Services). These articles demand reciprocity and the liberalisation of 'substantially all' tariffs over a 'reasonable period of time'.[2] Critics of the EU's position fall into two camps: one advocating 'alternative EPAs' and one pressing for 'alternatives to EPAs'. Both groups represent possible impediments to a successful conclusion of the strategy.

Despite the EU's emphasis on EPAs representing the 'only game in town' and subsequent reluctance to discuss alternatives, Cotonou's Article 37 does, in theory, allow for deviations from the main path:

> The Community will assess the situation of the non-LDCs which, after consultation with the Community, decide that they are not in a position to enter into economic partnership agreements and will examine all alternative possibilities, in order to provide these countries with a framework for trade which is equivalent to their existing situation and in conformity with WTO rules.

Table 9.1 ACP countries according to region (LDCs in Bold)

Region	Component states
Caribbean	Antigua and Barbuda, Bahamas, Barbados, Belize, Dominica, Dominican Republic, Grenada, Guyana, **Haiti**, Jamaica, St Kitts and Nevis, St Lucia, St Vincent and the Grenadines, Suriname, Trinidad and Tobago.
Central Africa	Cameroon, **Central African Republic**, **Chad**, Congo Republic, **Equatorial Guinea**, Gabon, **São Tomé** and **Príncipe**.
Eastern and Southern Africa	**Burundi**, **Comoros**, **Democratic Republic of Congo**, **Djibouti**, **Eritrea**, **Ethiopia**, **Kenya**, **Madagascar**, **Malawi**, Mauritius, **Rwanda**, Seychelles, **Sudan**, **Uganda**, **Zambia**, Zimbabwe.

Continued

Table 9.1 Continued

Region	Component states
Pacific	Cook Islands, Federation of Micronesia, Fiji, **Kiribati**, Marshall Islands, Nauru, Niue, Palau, Papua New Guinea, **Samoa, Solomon Islands,** Tonga, **Tuvalu, Vanuatu.**
SADC-minus	**Angola**, Botswana, **Lesotho, Mozambique,** Namibia, Swaziland, **Tanzania.**
West Africa	**Benin, Burkino Faso, Cape Verde,** Côte d'Ivoire, **Gambia**, Ghana, **Guinea, Guinea-Bissau, Liberia, Mali, Mauritania, Niger**, Nigeria, **Senegal, Sierra Leone, Togo.**

The EU model

While, theoretically, alternatives to EPAs exist, there are manifold reasons behind the EU's reluctance to reveal what these alternatives might entail or what form they might take. Critical factors include the erosion of the EU-ACP 'special relationship', the economic marginalisation of the ACP bloc, and the opening up of the 'near abroad' after the collapse of communism in Eastern Europe. However, EPAs fulfil a number of additional criteria for EU policymakers: WTO compatibility; the 'normalisation' of trade with the ACP; and a strengthening of the neoliberal logic of increased global trade liberalisation. Regionalism is also purportedly 'good' for developing countries as, effectively implemented, it results in economies of scale, reduced costs and improved governance, and affords improved levels of regional peace and security.

The EU itself provides an undeniably tempting end goal for those advocating regionalism as a vehicle for development. Europe, economically stagnant for much of the late 1960s and 1970s, having been squeezed out of the global economy by the producers in the USA and, more importantly, by aggressive Japanese producers, managed, through political and economic integration, to become the richest economic bloc within the global economy. As a result, a significant amount of theorising on the origins and nature of EU integration has taken place over the past half century, much of which has centred on the degree to which the EU represents a *sui generis*, a unique case or a potential template for other regions to follow.

Despite the longevity of integration theory, with David Mitrany (2003) putting forward his view of an alternative world order in 1943, it has not been without its detractors. Some critics have contended that it will amount to 'a rather long but not very prominent footnote in the intellectual history of twentieth century social science'(Donald J. Puchala cited in Chryssochoou 2001, p. 9). In certain respects, this represents a valid critique, particularly given the fluctuating fortunes of integration theory as a subfield within international relations. The majority of detractors point to the fact that integration theory is essentially confined to a single case study – that of the EU, and that it has no wider application. Moreover, no grand theory capable of accounting for this particular case has been able to stand up to sustained scrutiny. One solution to the dilemma has been for theorists to 'cobble together' various theoretical perspectives in order to arrive at a basic understanding of the integration process.[3] For many critics, such a step defeats the *raison d'etre* for theorising, given that such arguments offer only singular explanations for unique processes. Furthermore, such piecemeal theories offer no wider application for other regions and have little predictive value. This, then, is the basis for the argument, in some quarters, that developing countries have little to learn from the European experience, with Markus Jachtenfuchs (2001, p. 259) arguing that a 'theory of European integration' is neither 'feasible nor desirable' and that theory-driven debates are often 'sterile and decoupled from empirical reality'.

The integration debate has certainly seemed at times to be a somewhat impoverished one, given the absence of overarching grand theories or even vigorous competing views, but this was not always the case. In the 1940s and 1950s, integration theories such as federalism and functionalism tended to be broadly normative; prescriptive rather than descriptive and, in essence, 'grand theories'. Even with the advent of less normative theories such as neofunctionalism, the notion of grand theory was maintained and extrapolated to other parts of the globe. The 'father' of neofunctionalism, the American theorist Ernst Haas, focused his attention on a number of regional organisations that developed following the Second World War, including the West European Union, the European Coal and Steel Community and the Nordic Council. Haas sought to understand forms of global integration, with his study of the International Labour Organisation forming the basis for much of his later work (Hooghe 2001, p. 1). Other key neofunctionalists, such as Karl Deutsch, Philippe Schmitter and Joseph Nye, also believed their work to have a global application rather than being exclusively tied to the process of *European* integration (Hooghe 2001, p. 2).

However, neofunctionalism gradually imploded under the weight of its own contradictions and its apparent failure to explain away the so-called period of 'Eurosclerosis' in the 1970s. The fact that it limped on for so long demonstrates the paucity of alternative theories during this period. With the rise of intergovernmentalism and supranationalism in the 1980s, integration theorists abandoned any pretence of global application, becoming steadily more focused on the process of European integration. In attempting to explain the Single European Act (SEA) and Treaty on European Union (TEU), many scholars abandoned grand theory and strove instead to bring together various aspects of differing theories in order to adequately explain events. The rapid pace of integration since the signing of the TEU has reinvigorated the discipline, prompting competing theories[4] and enabling the field to progress beyond the confines of European integration.[5]

The one aspect of the EU integrationist debate that appears to have potential for wider application is that of 'spillover', a concept originally associated with the neofunctionalists. The dynamic for regional integration, according to Ernst Haas, is the initial merging of economically vital but politically unimportant sectors of the economy, overseen by an independent 'higher authority'. In the case of the EU, this merging is best demonstrated by the formation of the European Coal and Steel Community (ECSC) in 1951. Neofunctionalism argues that the result of such cross-border integration is increased trade between the countries involved, which in turn places pressure on related sectors to integrate in a similar fashion. Similarly, the process will allow interest groups traditionally confined to the national level, such as trade unions, to expand to the supranational level. People, becoming aware of a shift in power, will gradually begin to transfer their loyalty to the emergent institution. This in turn will create further pressure for increased integration. Proponents of neofunctionalism stress that greater economic integration will inevitably lead to increased political integration, almost as a side effect (Mutimer 1994, p. 29). This process of deepening integration, as one sector puts pressure on another to merge, is described as 'spillover'. Pressure to integrate in one area 'spills over' into other sectors and thereby serves to create further pressure. Leon Lindberg, another noted neofunctionalist scholar, defines spillover as 'a situation in which a given action, related to a specific goal, creates a situation in which the original goal can be assured only by taking further actions, which in turn creates a further condition and need for action and so forth' (cited in Rosamond 2000, p. 60). In basic terms, then, spillover can be used to explain the transition from the ECSC to Monetary Union and the Euro.

Yet, despite the appeal of the concept, by the mid-1960s it was apparent that neofunctionalism was flawed. The Community had come to an almost complete standstill and the idea of integration as an inexorable process of spillover was unable to account for the *impasse*. This new mood was most clearly epitomised by the actions of the French under Charles de Gaulle during the 'empty chair' crisis of 1965.[6] As a consequence of burgeoning French nationalism, the Community languished in a virtual state of inertia from the mid-1960s to the mid-1980s, despite the successive expansions in 1973 and 1980. Neofunctionalism was unable to account for this slowdown and in 1975 Haas himself declared the theory 'obsolescent' (cited in Groom 2003, p. 111). Furthermore, despite the rash of regional initiatives in the South in the 1960s and 1970s, the consequences were unexceptional, implying once again that the EU is a unique case.

However, spillover, as a concept, enjoyed a renaissance in the late 1980s and early 1990s, reintroduced by theorists such Robert Keohane and Stanley Hoffman (1990), and 'old school' neofunctionalists such as Philip C Schmitter (1996, 2004), as they sought to explain the European *relance*. What this analysis does appear to suggest is that, with a degree of political will, which has generally been perceived to be lacking in the South, regional economic integration initiatives rapidly become 'locked in', forcing further changes and thereby cementing reforms. However, the political will demonstrated by Thatcher, Mitterrand and Kohl in the build-up to the signing of the SEA in 1985, which served to reinvigorate the EU, remains notably lacking amongst many ACP political elites.

The ACP perspective

Despite the possible template suggested by the EU experience, EPAs pose a number of political and economic challenges for ACP countries. At the most basic level, the lowering of trade barriers will adversely affect revenue flows for a number of countries which rely heavily on duties to bolster state coffers (see Table 9.2). For the Ivory Coast, Sierra Leone and Uganda, the loss of this income could represent a serious financial blow: duties represent between 40and 50 per cent of total government revenue (Stop EPAs Coalition 2004, p. 5). Cuts in government spending and associated welfare losses are thus all but inevitable. Many states are therefore extremely reticent about ceding their tariff revenues in favour of a freer trade strategy aimed at, amongst other aspects, encouraging greater intra-regional trade (Narlikar 2005, pp. 1–3). Even allowing for the fact that these losses may only be short to medium term, it is difficult to stress the disquiet posed by this potential loss of income.

Table 9.2 Projected loss of revenue post-EPA

Country	**Loss of tariff/duty revenue post-EPA (American dollars)**
Angola	103,254,613.00
Benin	39,523,104.00
Botswana*	5,232,995.00
Burkina Faso	22,003,937.50
Burundi	7,664,911.00
Cameroon	149,256,117.00
Central African Republic	5,844,950.00
Chad	26,677,028.00
Congo Republic	75,104,052.00
Cote d'Ivoire	112,236,538.00
Djibouti	37,523,124.00
DRC	24,691,828.00
Equatorial Guinea	33,914,150.00
Ethiopia	55,126,359.00
Eritrea	7,385,208.00
Gabon	74,302,297.00
Ghana	193,683,365.00
Guinea-Bissau	1,990,216.50
Kenya	107,281,328.00
Lesotho*	256,314.00
Madagascar	7,711,790.00
Malawi	7,090,310.00
Mali	33,141,747.00
Mauritania	14,572,779.00
Mauritius	71,117,968.00
Mozambique	7,640,140.00
Namibia*	3,831,993.00
Niger	20,487,214.00
Nigeria	426,902,557.50
Rwanda	5,622,946.00
Senegal	80,203,188.50
Seychelles	24,897,374.00
Sudan	73,197,468.00
Swaziland*	811,140.00
Tanzania	32,490,659.00
Togo	35,471,728.00
Zimbabwe	18,430,590.00

* SACU members (Karingi *et al.* 2005).

Furthermore, the lowering of trade barriers will expose domestic producers to increased levels of competition from neighbouring firms and from abroad. ACP countries are also concerned that, given their restricted institutional capacity, they are under-equipped to extract significant concessions from the EU. The EU has argued that it will be prepared to negotiate EPAs with regional economic communities (RECs) comprising a free trade area (FTA) or customs union (CU). The regional aspect, once again, poses a number of challenges. Historically, regionalism has not fared well in the South. Africa, in particular, has an extremely poor record. Intra-African trade is notoriously weak. Within the Southern African Development Community (SADC), for instance, intra-regional exchange, excluding trade with South Africa, accounts for a mere 4 per cent of overall trade (Jenkins 2001, p. 20). Similarly, in West Africa, such trade accounts for just 13 per cent of total trade (Bilal *et al.* 2006b). This is largely since, unlike pre-EU patterns of trade in Europe, the majority of trade flows between ACP countries take the form of 'inter-industry' rather than 'intra-industry' flows. In essence, the reason EU integration was successful was because there was a reciprocal demand for imports amongst the relevant states. States engaged in a complementary exchange of similar goods and services (intra-industry trade). The result of such interaction is a form of integration that is mutually agreeable and does not result in the large-scale displacement of domestic concerns. Rather than a zero-sum game, it is a relationship that is mutually beneficial. Between developing countries, the pattern tends to be different (inter-industry trade). Given that the majority of ACP countries in any given region tend to produce similar goods for export, such as raw materials, agricultural goods, and 'semi-manufactures', competition tends to be based solely on price structures rather than any form of product differentiation. This can often lead to acrimonious trade relations as countries are overcome by their mercantilist instincts. Economic modelling by World Bank analysts Jeffrey Lewis *et al.* (2002, p. 16) suggests that the SADC countries stand to gain far more from unilateral agreements with the EU than through a regional FTA. In nearly all cases, access to EU markets provides for substantially greater gains than would a fully liberalised regional bloc.

An examination of export dependence on the EU on the part of the various SADC members provides a clear illustration of the weakness of the South African economy where the boosting of regional growth is concerned. According to Lewis *et al.* (2002, p. 15), Botswana directs 76 per cent of its exports towards the EU whilst Malawi and Mozambique rely on European markets to absorb approximately 40 per cent of their exports. South Africa, whilst it is Africa's strongest economy, attracts little trade from its neighbours. Zimbabwe – before the escalation of political and economic instability – was its strongest trading partner in the region, yet South Africa received just 14 per cent of Zimbabwean exports. Meanwhile, Zambia, Mozambique, Tanzania and Malawi all direct less than 10 per cent of their total exports to South African markets (Lewis *et al.* 2002, p. 17). The benefits of FTAs, for many ACP countries, are thus to be adequately demonstrated. This scepticism has been reinforced by United Nations Economic Commission for Africa (2005a, p. 27; 2005b, pp. 18–20) calculations which suggest that EPAs will bring about a reduction in intra-regional trade in favour of EU exporters.

A lack of political will, alluded to above in economic terms, is frequently based on enmity towards established regional rivals. Once again, it is in Africa where these problems are most apparent. The rivalry between Zimbabwe and South Africa in SADC has, at times, almost descended into farce. Before the end of *apartheid,* President Robert Mugabe of Zimbabwe had been viewed as the leader and elder statesman of the so-called 'front-line states' and had achieved a relatively high international profile. This was to change when Nelson Mandela acceded to power in South Africa in 1994, eclipsing Mugabe as the region's most eminent political figure. The adjustment sparked an intense competition for the control of SADC that threatened to undermine the entire Community (Simon and Johnson 1999, p. 2). Similarly, the intense competition between Cameroon and Gabon has undermined attempts at integration in the Central African Communauté Economique et Monétaire de l'Afrique Centrale (CEMAC; Zafar and Kubota 2003, p. 17). Chad, too, has a history of conflict with its neighbours. In West Africa, Anglophone/Francophone splits have rendered cooperation almost impossible at times, with further complications arising through currency incompatibility between CFA zone and non-CFA zone countries (Page and Bilal 2001, p. 17). Tanzania, Uganda and Kenya have struggled to cooperate as members of the East African Community. In the Pacific region, Fiji and Vanuatu have been at loggerheads over trade issues[7] and Fiji, as the largest actor in the group, is frequently viewed with suspicion by its regional partners. The Caribbean group, whilst less fractious, is divided along former colonial lines. Given the time constraints placed on ACP countries by the EPA schedule, it is difficult to see how these hurdles will be satisfactorily overcome.

Moreover, the fact that LDCs are entitled to be treated as 'special cases', by virtue of the Everything But Arms (EBA) initiative (see Chapter 4), has created further obstacles to the formation of EPAs.[8] Least Devoloped

Countries have little to gain from signing up to EPAs as these will effectively nullify their strategic position vis-à-vis the EU which, as of 2009, will allow all LDC exports to enter the Polity tariff- and quota-free. More importantly, the EBA initiative does not require reciprocity. Under the regime, LDCs will be allowed to continue to raise revenues from tariffs. As highlighted at the beginning of the chapter (Table 9.1), each regional negotiating group contains a number of LDCs and their cooperation will be essential if the EPAs are to become viable. It is only because EBA does not contain an aid element and, furthermore, is not enshrined in treaty form, that LDCs will probably be convinced to partake in the EPA process.

The Doha Round and the shape of EPAs

As mentioned, the EU has set great store by the compatibility of EPAs with WTO rules. While seemingly straightforward in theory, this demand is problematic for two reasons. The first is that the section governing FTAs, Article XXIV of the GATT, whilst offering guidelines, is open to interpretation where specific application of the principles embodied therein are concerned. For example, paragraph 8 (b) states

> A free-trade area shall be understood to mean a group of two or more customs territories in which the duties and other restrictive regulations of commerce are eliminated on *substantially all the trade* between the constituent territories in products originating in such territories.

However, nowhere is it specified what 'substantially all the trade' entails. This creates a challenge for policymakers. Trade liberalisation can be measured in a number of ways including volume, percentage, value or tariff line (Bilal and Rampa 2006). The EU's interpretation is that 'substantially all' equates with a liberalisation of 90 per cent of the 'total value of trade' between the contracting parties. Moreover, the EU envisages this figure as an average between the Polity and various ACP groupings. In other words, the EU interpretation allows for an asymmetrical approach whereby one party might liberalise more extensively than the other, provided that the mean is 90 per cent. In the case of the Trade, Development and Cooperation Agreement (TDCA) signed between the EU and South Africa in 1999, the Polity agreed to a ratio of 95:86 per cent, thereby crossing the 90 per cent threshold. Similarly, the 'reasonable length of time' stipulated in Article XXIV is not specified. The EU has interpreted this as a period of ten years. However, precedent suggests that between 12 and 20 years is more practical (Bilal and Rampa 2006). If challenged by a third party, the WTO might conceivably rule against this interpretation.

The second key difficulty pertaining to EPAs is that the rules governing regionalism form part of the ongoing Doha negotiations and might well be subject to change on the completion of the Round. The Doha Round was scheduled for completion by January 2005, well before the EPA deadline of 31 December 2007. However, the Doha Round has stalled and might yet drag on until 2009. The EU is thus demanding that the EPAs be in place before the rules governing such structures are clarified. The implications are significant. Article XXIV of the GATT was designed with developed countries in mind and thus does not offer flexibility in terms of levels of reciprocity. The ACP bloc submitted a proposal to the WTO in 2004 demanding that a 'special and deferential' (SDT) clause be inserted in Article XXIV (ACP Group of States 2004). Support for this motion amongst the WTO member states has been mixed.

EPAs are thus technically 'jumping the gun' somewhat and might well be open to challenges within the WTO once the Doha Round has been completed, given that they are largely based on the EU's interpretation of prevailing WTO rules. However, since the WTO waiver for the preferences afforded the ACP bloc by the EU will expire at the beginning of 2008, failure to agree on EPAs might well result in many ACP countries losing their privileged access to EU markets, which could be potentially financially ruinous for many exporters. The horticulture sector in Kenya, for example, has expanded dramatically, to the extent that Kenya now supplies over 50 per cent of all roses imported into the EU (Krimphoff 2007, p. 58). The industry employs approximately 135,000 workers (Stevens 2007, p. 2) and studies have drawn a link between horticultural exports and poverty alleviation (McCulloch and Ota 2002). The margins on commodities such as cut flowers are small and the sector is profitable largely as a result of Kenya's privileged access to EU markets. Without such preferential treatment, Kenyan flowers would have to compete with those of more efficient competitors in Latin America, with predictable results. Similarly, Namibian beef exports to the EU would also suffer greatly if preferential access was denied (Stevens 2007, p. 2).

Eroding regional initiatives

In certain quarters, there has been concern that the EPAs will actually undermine established regional integration efforts, particularly amongst the African ACP members. The most notable example is that of SADC. One of the longer established regional groupings, the forerunner of SADC, the Southern African Development Coordination Conference (SADCC), was established in 1980. The Community, encompassing 15 member states, has agreed on a host of integration measures including a FTA by 2008, a customs union by 2010, a common market by 2012 and a central bank and single currency by 2016. Although progress has been slow, SADC stands to be wholly undermined by the EPA process. First, because South Africa, the regional hegemon, was never afforded full ACP status, it was precluded from signing an EPA as part of the wider ACP and instead negotiated its own bilateral treaty with the EU – the EU-SA TCDA. However, South Africa is a member of the Southern African Customs Union (SACU) with Botswana, Lesotho, Namibia and Swaziland (BLNS), which has a common external tariff. The result is that these four countries are *de facto* signatories to the EU-SA agreement. Consequently, five of the SADC member states are, in effect, already part of an EPA with the EU. As stated above, a number of African ACP members hold overlapping regional memberships and the EU negotiations have forced them to choose which configuration they wish to be part of during this process. Only seven member states opted to negotiate a SADC–EU EPA. The remaining SADC members defected to the specifically established Eastern and Southern Africa (ESA) configuration which compromises members drawn from SADC, the Common Market for Eastern and Southern Africa (COMESA), the East African Community (EAC) and the Indian Ocean Commission (IOC).

Economic Partnership Agreements can encompass FTAs or customs unions. Given the basic requirement of a common external tariff, a state cannot belong to two customs unions. If the SADC members of the ESA configuration were to decide on such an EPA strategy, the SADC roadmap would be destroyed. Similarly, it would clash with the already established EU–SA Agreement. Additional complications arise as a result of the EAC Customs Union, established in 2005, and the fact that the COMESA members are working towards the creation of a customs union by 2008. With regards to the EAC Customs Union, Kenya has chosen to negotiate an EPA as part of the ESA configuration whilst Tanzania is attempting the same under the SADC umbrella. There has been little guidance from the EU as to how such issues might be resolved, particularly given the fact that it is unclear whether, on the completion of negotiations, states will enter into agreement with the Polity individually or as part of a regional collective. At the time of writing, even at this very late stage of the negotiations, the regional signatories of the treaty remain to be designated.

The 2008 deadline

Much depends on the readiness of the regional groupings to speed up the process. Of the African groups, *West Africa* (ECOWAS) is arguably at the most advanced stage. It has held two rounds of negotiations with the EU and has presented a draft EPA structure to the Commission's negotiators who, in turn, have presented a draft EPA text for ECOWAS' consideration. Both sides have agreed that these two documents will guide the final EPA text (Julian 2007, pp. 10–11). *Central Africa* (CEMAC) and the Commission have agreed a draft structure of the eventual EPA. However, negotiations have largely stalled on the issue of political conditionality, together with demands for greater compensation for estimated adjustment costs (European Centre for Development Policy Management 2006b). Negotiations between the Commission and *Southern Africa* (SADC) have progressed slowly due to the complex nature of the grouping. As stated, South Africa has observer status only. Botswana, Lesotho, Namibia and Swaziland, as members of the SACU, are using the EU–SA TCDA as the basis of their offer to the EU. The remaining SADC members; Angola, Mozambique and Tanzania, are developing their own demands based on the EBA initiative (European Centre for Development Policy Management 2006e). The *Eastern and Southern Africa* (ESA) configuration and the EU agreed a 'roadmap for negotiations' in 2004 and the region has duly progressed with the dialogue. Both sides have approved a joint agreement on trade in goods and are proceeding with a draft EPA text. Difficulties remain regarding issues such as market access, and financial and development costs (European Centre for Development Policy Management 2006d; Julian 2007, pp. 10–11). The *Caribbean* (CARIFORUM) negotiators agreed upon a 'roadmap' for discussions and settled on a timetable for the implementation of stated goals. Both parties have agreed to the basic principles of the EPA and, although impediments remain, CARIFORUM is the closest of the six to finalising its negotiations with the Commission (European Centre for Development Policy Management 2006a; Julian 2007, pp. 10–11). Finally, the *Pacific* region (PACP) has submitted a draft EPA to the Commission for its consideration.

However, PACP has been forceful in its demands for greater funding for EPAs and this issue has become something of a sticking point, capable of derailing negotiations (European Centre for Development Policy Management 2006c).

Much has been made of the 31 December 2007 deadline for the signing of the EPAs. As discussed above, the WTO-sanctioned waiver for a continuation of the Lomé preferences expires on the same date. However, this deadline is not as black-and-white as it would first appear and there is a degree of 'wriggle-room' for those hoping to avoid a summary loss of trade preferences on 1 January 2008. A complaint would have to be filed against the ACP preferences within the WTO in order to bring WTO dispute settlement procedures into play. The number of complaints lodged with the WTO peaked in 1997 and have subsequently declined dramatically (Leitner and Lester 2004, p. 171). The procedures are notoriously slow and potential complainants would only be likely to act if the ACP appeared to be negotiating in 'bad faith'. This state of affairs could conceivably give the ACP countries an additional two-year window within which to conclude EPA negotiations (Stevens 2006, p. 456, n3). This period could conceivably be extended further if the USA–Africa Growth and Opportunity Act provides a comparable precedent. This legislation, enacted in 2000 and offering preferential access to American markets for African goods, is set to run until 2015 despite never being afforded a waiver. The Act is yet to be challenged in the WTO. The EPA negotiations are therefore not necessarily under threat if the agreements have not been successfully concluded by 31 December 2007.

Conclusion

The EU is determined to press ahead with its vision of a post-Lomé partnership agreement with the ACP. Central to this vision is the negotiation of EPAs between the EU and six regional ACP clusters. These partnership agreements will be WTO-compatible, and conform to the EU's strongly neoliberal approach to development and international trade. However, many ACP countries are deeply concerned about the prospect of the increased regional integration and trade reciprocity that is inherent within the new regime. Obstacles to successful regional FTAs abound: historical rivalries, mercantilist tendencies, inter-industry versus intra-industry trade and a lack of political will. If successful, however, the EPAs could potentially provide significant benefits: economies of scale, reduced transaction costs and good governance, together with peace and stability. The success of the EPAs is largely dependent on the degree to which the integration initiatives become imbedded. It is thus crucial, if EPAs are to be successful, that the participants come to 'own' the process. At present it appears that the majority of ACP states, rather than embracing the strategy, are having EPAs thrust upon them.

Conclusion

The Cotonou Agreement has, at its core, three stated goals; the fostering of poverty alleviation, sustainable development and the gradual of the ACP economies into the global economy. It is on the basis of these three criteria that the implementation of the Agreement must be judged. This study has therefore focused on the EU's commitment to the ACP countries in this regard. The conclusion, based on the evidence presented here, is that the EU has failed to embrace the 'spirit' of the Agreement and has opted instead for a rigid, 'ideological' development framework that is frequently either inflexible or inappropriate to the situation concerned, or both. Moreover, the environmental aspects of 'sustainable development' are not always afforded sufficient priority.

Poverty and environmental degradation feed off each other. High levels of poverty force people to exploit their surroundings in an unsustainable manner in order to survive. In the rural areas of developing countries this can result in soil erosion, deforestation, desertification and biodiversity loss, all of which in turn serve to diminish productivity. Poverty reduction and environmental degradation are thus two sides of the same coin. If the Millennium Development Goal to halve the poverty by 2015 is to be achieved, then those living in extreme poverty, mostly in rural areas, must be targeted. The potential for improved livelihoods, incorporating reductions in rates of infant mortality and disease as well as bolstered food security, literacy levels and wealth, is inextricably tied to environmental protection. Since environmental concerns play such an immediate role in determining people's well-being, it is imperative that any development policies implemented by bodies such as the EU address 'green' issues as part-and-parcel of any wider development strategy. If the EU ignores the importance of environmental issues to the development debate, or pays them only nominal consideration, this will effectively cripple any real prospect of meeting MDG targets.

As has been demonstrated, the EU has publicly embraced sustainable development as a guiding principle for both domestic and international policy, and thus appears committed to an environmentally proactive development agenda. Through analysis of relations with the ACP, a group comprising many of the poorest countries in the world, it is ossible to argue that the EU has been relatively unsuccessful where the application of this principle has been concerned. This study posits the need for the EU to reconsider its 'one-size-fits-all' approach to development and to take into account the contention that globalisation does not necessarily require the abandonment of all but market-driven strategies.

For those who agree that the earth has a limited 'carrying capacity' and that this 'Plimsoll line' (see Daly's argument, Chapter 2) is in danger of being overwhelmed, the EU's approach to sustainable development can only be viewed as restricted. However, a radical agenda centred on a policy of zero-growth and qualitative refinement would be at odds with the EU's perception of the extent of the current global environmental crisis and how this can be best addressed. A sustainable development strategy based largely upon market-driven responses to environmental degradation is a hallmark of EU development policy and characterises the Polity's approach to its relations with its ACP partners. This particular strand of sustainable development maintains that 'getting prices right', technological innovation and the generation of wealth will curb the worst excesses of environmental degradation, especially those associated with high levels of poverty. It is also fully compatible with prevailing neoliberal trends. The application of this approach by the EU is clear, in both macro and microeconomic terms.

The market-based approach to ensuring sustainable development through better accounting procedures and increased levels of wealth is, however, an argument that remains to be won. While certain pollutants do begin to decline in the face of increasing levels of wealth, this is not the case for all environmental indicators, especially those that are transboundary in nature. It is certainly arguable that an increase in overall wealth would serve to combat the environmental affects usually associated with extreme levels of poverty. Given that the available evidence does seem to suggest that economic liberalisation tends to bolster growth, there is some validity to the EU's approach to promoting sustainable development amongst the ACP countries by means of a focus on trade liberalisation and integration into the global marketplace. However, it can also be argued that even if the EU is correct in this regard, and that a weak sustainability approach does represent the best prospect for economic growth, poverty alleviation and environmental protection amongst the ACP countries, the Polity is often at odds with its own stated aims. Furthermore, it can be maintained that the EU's commitment to market-based solutions is often unrealistic and counterproductive where assessing the needs of the poor is concerned. Market-based strategies can only be effective in sectors where genuine markets can exist and where there is an ability to pay. The EU's commitment to water privatisation in the South is a case in point, given that, in many instances, associated price rises have forced the poor to consume from unsafe sources. One outcome of this has been a resurgence of supposedly nineteenth-century epidemics such as cholera. The adherence to private sector ventures is at variance with the strategies adopted by the most successful programmes designed to provide basic services. Botswana's publicly owned Water Utility Corporation (WUC) increased access from 30,000 in 1970 to 330,000 in 1998 while the mass electrification programme in South Africa effectively doubled the number of homes connected to the national grid in under a decade.

For the EU, the key to facilitating the perceived benefits of the new Cotonou Agreement is that of regionalism. With the demise of Lomé, the EU was determined that the ACP countries band together in regional groupings in order to negotiate economic partnership agreements with the Polity based on neoliberal principles, thereby destroying the previous system of preferential, non-reciprocal access to EU markets for ACP goods. It is argued by proponents that this increased regionalism will boost exports, thereby encouraging economic growth, creating markets of scale and aiding a reduction in poverty levels.

In addition, the Cotonou Agreement's stipulations have effectively removed the 'collaborative' aspects of the ACP-EU relationship and have served only to emphasise the David versus Goliath nature of the association. Whereas ACP countries were able to negotiate with the EU from a position of strength in the mid-1970s, by the time the Lomé IV-bis expired in 2000 the influence of the ACP bloc had waned significantly. This disparity has resulted in a resurgence of the dependency-style language used to describe the relationship between the two and has brought the North–South debate sharply into focus. Accusations of neo-imperialism/neo-colonialism were levelled against the EU by a number of African leaders in the lead-up to the signing of the Agreement. The acrimonious nature of the negotiations between South Africa and the EU before the signing of the joint was a foretaste of things to come and made plain the Polity's determination to impose its will on the shaping of all future agreements.

However, if, despite such difficulties, the EU is correct in its assumption that its market-based approach to sustainable development represents the best way forward, then the EU is well placed to play a positive role in promoting related ideals. The EU has contributed significant levels of funding with such a strategy in mind; any evidence of success must be gauged in terms of progress made in fulfilling the MDG of halving the number of people living in extreme poverty by 2015. In order to assess the efficacy of EU funding relating to global market integration, poverty and sustainable development, this study has concentrated on four specific sectors: trade, agriculture, energy provision and water and sanitation. The evidence demonstrates that many of the policies and strategies adopted by the EU in order to effect its development vision have not, at least from an ACP perspective, had the desired impact. This 'deficit' can largely be attributed to the EU's 'one-size-fits-all' development strategy.

When dealing with the plight of those living in extreme poverty, market-based initiatives are seldom effective at ensuring fair levels of distribution. As shown, South Africa has achieved great success with its electrification programme precisely because the project was funded and planned by the public sector. There have been few notable successes in expanding access to basic utilities such as water and sanitation when these industries have been privatised – the latter being a strategy heavily promoted by the EU. Examples from South Africa and Tanzania suggest that the privatisation of such sectors leads to higher prices, which in turn can have a devastating effect on poor communities, the cholera outbreak in South Africa in 2000, precipitated by higher water prices in KwaZulu-Natal, being a case in point. Denying people access to basic services such as clean, safe drinking water will always have a negative impact on poverty indicators. The irony of the apartheid government in South Africa being more effective in this regard speaks volumes. Yet, despite such examples of private sector failure, the EU appears to remain committed to this approach to the provision of basic services.

If liberalisation and trade are to be the cornerstones of poverty-reduction strategies, then the EU must practice what it preaches. There appears to be little rationale behind the encouragement of developing countries to liberalise their markets and to boost exports if any competitive advantage they secure by doing so is nullified by EU domestic policies. The Mid-Term Review of the CAP, with its shift away from 'blue' to 'green' box subsidies, while more WTO 'friendly', does not address the imbalances perpetuated by the regime. Single farm payments, whilst no longer coupled to production, still have a trade-distorting effect. Evidence from the USA as well as from the EU demonstrates that subsidies, whether coupled to production or not, quash the comparative advantage held by developing countries in the agricultural sector. To argue that because such subsidies are now green box they have no trade-distorting effects is disingenuous in the extreme. The continuing subsidisation of the average EU farmer to the effect of $16,000 a year makes it difficult to see how ACP countries will be able to compete with their European counterparts. It was agreed at the conclusion of the Uruguay Round that developed countries would reduce agricultural subsidies by a minimum of 20 per cent by 2000. Instead, the corresponding period witnessed an increase in agricultural support in the North from $182 billion to $300 billion. There is thus little evidence to suggest that any of the developed nations are taking seriously this commitment to reduce subsidies – as the failure of the WTO summits in Seattle, Cancun and Hong Kong testify. More importantly, the MTR of the CAP does not address questions of market access; this particular issue will only be addressed in 2013, when the next full review of the regime is scheduled.

Perversely, the reforms enabled by the MTR could conceivably disadvantage ACP countries. The high domestic prices generated by the CAP regime have been of benefit to ACPs, enabling producers to sell their agricultural products at prices that are frequently three times higher than the global average. Given that the most likely effect of the MTR reforms will be to drive down domestic prices in Europe, many ACP countries will be faced with the prospect of earning less for their exports whilst being unable to increase the volume of these (the exception being the LDCs, by virtue of the Everything But Arms Initiative, but only if they prove capable of dramatically bolstering their productivity levels).

This is not to argue that Cotonou is without value. There are numerous EU-funded projects and programmes that demonstrate enormous potential where poverty alleviation and sustainable development are concerned. However, the majority of promising projects have tended to be small-scale, localised projects such as the 'Thousand Schools' electrification programme; essentially pilot studies or 'one-offs'. As such, these do not represent development templates. All of the large-scale initiatives, such as EBA, the EUEI and the EUWI, place a heavy emphasis on market-based solutions to the problems of poverty and environmental degradation. Based on the evidence presented, it is difficult to see how such strategies will improve these indicators. Given that those living in extreme poverty are seldom in a position to take advantage of the opportunities afforded them by increased access to the global economy, markets of scale or privatisation, it is imperative that any meaningful attempts to alleviate their plight look beyond basic market strategies. Efficient markets may allocate resources optimally but they rarely do so based on need.

Appendices

Appendix I

The African, Caribbean and Pacific Countries

Africa

- Angola
- Benin
- Botswana
- Burkina Faso
- Cameroon
- Cape Verde
- Central African Republic
- Chad
- Comoros
- Congo (Brazzaville)
- Cote d'Ivoire
- Democratic Republic of Congo
- Djibouti
- Equatorial Guinea
- Eritrea
- Ethiopia
- Gabon
- Gambia
- Ghana
- Guinea
- Guinea Bissau
- Kenya
- Lesotho
- Liberia
- Madagascar
- Malawi
- Mali
- Mauritania
- Mauritius
- Mozambique
- Namibia
- Niger
- Nigeria
- Rwanda
- Sao Tome & Principe
- Senegal
- Seychelles
- Sierra Leone
- Somalia
- South Africa
- Sudan
- Swaziland
- Tanzania
- Togo
- Uganda
- Zambia
- Zimbabwe

Caribbean

- Antigua
- Bahamas
- Barbados
- Belize
- Cuba
- Dominica
- Dominican Republic
- Grenada
- Guyana
- Haiti
- Jamaica
- St Kitts & Nevis
- St Lucia
- St Vincent
- Suriname
- Trinidad & Tobago

Pacific

- Cook Islands
- East Timor
- Federated States of Micronesia
- Fiji
- Marshall Islands
- Nauru
- Niue
- Palau
- Guinea
- Samoa
- Solomon Islands
- Tonga
- Tuvalu
- Vanuatu

Appendix II

Least developed countries

Country	GDP per capita (in $) (2002–2004 figures)
Afghanistan	700
Angola	1,900
Bangladesh	1,900
Benin	1,100
Bhutan	1,300
Burkina Faso	1,100
Burundi	4,600
Cambodia	1,900
Cape Verde	1,400
Central African Republic	1,100
Chad	1,200
Comoros	700
Dem. Republic of Congo	700
Djibouti	1,300
Equatorial Guinea	2,700
Eritrea	700
Ethiopia	700
Gambia	1,700
Guinea	2,100
Guinea Bissau	900
Haiti	1,600
Kiribati	800
Lao PDR	1,700
Lesotho	3,000
Liberia	1,000
Madagascar	800
Malawi	600
Maldives	3,900
Mali	900
Mauritania	1,800
Mozambique	1,200
Myanmar	1,800
Nepal	1,400
Niger	800
Rwanda	1,300
Samoa	5,600
Sao Tome & Principe	1,200
Senegal	1,600
Sierra Leone	500
Solomon Islands	1,700
Somalia	500
Sudan	1,900
Tanzania	600
Togo	1,500
Timor-Leste	500
Tuvalu	1,100
Uganda	1,400
Vanuatu	2,900
Yemen	800
Zambia	800

The UN reviews the status of LDCs every three years. In order to qualify as an LDC, a country must demonstrate the following features:

- A gross national income per capita of less than $750.
- A low Augmented Physical Quality of Life Index (which is calculated on indicators for nutrition, health, education and adult literacy).
- Economic vulnerability as verified by a poor Economic Vulnerability Index (based on indicators such as agricultural sector instability, exports of goods and services, importance of non-traditional economic activities to the economy and the impact of natural disasters on the population).

To graduate from the list, a country must meet the basic levels for at least two of the three criteria outlined above for two successive reviews. Senegal was added to the list in 2000 and Timor-Leste in 2003, bringing the total to 50.

Notes

Introduction

1. See Appendix I.
2. The EU itself defines poverty as 'not simply as the lack of income and resources, but also encompassing the notion of vulnerability and such factors as no access to adequate food supplies, education, health, natural resources, drinking water, land, employment and credit facilities information and political involvement, services and infrastructure' (European Centre for Policy Development 2002, p. 2).
3. In southern Africa, reliance on non-agricultural sources of income for the rural poor is estimated to be as high as 80–90 per cent of household earnings (Chisholm 2001, p. 12).
4. In terms of food security, on a national level, self-sufficiency entails 'meeting food needs as far as possible from domestic supplies and minimising dependence on food trade' whilst the concept of self-reliance 'takes into account the possibilities offered by international trade' (UNESCAP 2004–2005).
5. Deep Ecology argues that all aspects of nature have intrinsic worth regardless of the value placed upon them in terms of their use to humans. Diversity represents value in its own terms and should only be impacted upon in order to serve vital needs. In order to preserve diversity we must, as a species, re-evaluate our relationship with nature and move away from destructive economic patterns which result in extinction and pollution. Reform of the current system is thus not an option. A complete reordering of priorities is required.

1 The EU and the ACP – from Yaoundé to Cotonou

1. The European Unit of Account (EUA), introduced to calculate balances in the European Payments Union (later the Organisation of Economic Cooperation and Development) in 1950 and pegged at the level of the Dollar, was the forerunner of the European Currency Unit (ECU), itself a forerunner of the Euro.
2. STABEX was introduced as part of the Lomé I regime and involved compensation to ACP member states if agricultural commodity prices fell below expected levels. SYSMIN, introduced after negotiations leading to Lomé II, performed a similar function for countries exporting minerals. Both these programmes were funded by the European Development Fund (EDF).

2 Towards an EU Definition of Sustainable Development

1. The Club of Rome is a 'think-tank' founded in Rome in 1968 in an effort to broaden what the founders perceived to be the overly narrow debate surrounding global issues, including issues pertaining to the environment.
2. The concept of 'natural capital' is a relatively new one. In orthodox economic theory, 'capital' generally referred to goods produced by human hands. Land, for example, was not viewed as capital. However, the growing concern surrounding environmental issues that developed in the 1970s and 1980s made certain economists re-evaluate this narrow definition. The environment has a 'productive capacity' and can be affected by human intervention and thus it made sense for economists to refer to 'natural capital'.
3. Ehrlich argued that as natural resources became increasingly scarce, so the price for these commodities would rise. In 1980, Simon offered Ehrlich a wager of $1000 that market prices for metals a decade later would not indicate declining stocks. Ehrlich was allowed to choose quantities of five metals which together totalled $1000 at then current market prices (he chose copper, chrome, nickel, tin and tungsten). It was agreed that if by 1990 the value of these metals was worth more than $1000 (after adjusting for inflation), he would win the bet. In the event that the metals were worth less than $1000, Simon would win the wager. The result was a vindication for Simon, with the collapse of the price of tin being the most dramatic of the five metals; a fall of 72 per cent. The 1990 value of the metals was $576.07 (after adjustments) (Tierney 1990).
4. Neoliberal environmental economics broadly advocates the use of market-based solutions for environmental problems and places particular emphasis on 'getting prices right' through the incorporation of the full costs of economic activity into the final price of any given commodity. David Pearce became a favourite of Margaret Thatcher when environmental concerns began to catch the imagination of the public in the mid- to late 1980s.
5. Hakonsen and Mathiesen (cited in Cuervo and Gandhi 1998, p. 36) argue that, if implemented in Norway, a carbon dioxide tax resulting in 20 per cent cuts could lead to a 1 per cent rise in GDP. Similarly, Welsch (cited in Cuervo and Gandhi 1998, p. 39) concludes that carbon/energy taxes can lead to increased employment if revenue is spent effectively.
6. The term implies that the earth is a 'closed system' and that consequently its resources are finite. If this is the case, then economists must take into account issues such as limits to growth.
7. Entropy is a measure of the availability of energy. Coal has a low entropy value; its energy is 'stored' and is available for use. Once the coal has been burned, this energy achieves a high entropy value as the process results in heat, which dissipates. Once dissipated, this energy is no longer as readily available as it was when it was contained within the coal (Jacobs 1991, p. 12).
8. Nicholas Georgescu-Roegen introduced the concept of applying the Laws of Thermodynamics to economic theory in 1971. The First Law of Thermodynamics states that energy is neither created nor destroyed but merely changes form, while the Second Law states that, in entropy terms, the cost of any biological

or economic activity is always greater than the resulting product. Economic growth is thus constrained by limited supplies of low entropy resources (Jackson 1996, pp. 8–12).

3 The Rise of the Neoliberal Consensus

1. Technically, making reference to a 'Marshall Plan' for Africa is historically misleading given that this aid was for the *reconstruction* of Europe ante-1939. Industrial capitalism and its related values were already well entrenched. The Marshall Plan was thus not a development strategy. Despite this, calls for a 'Marshall Plan for Africa' are made on a regular basis (see Mbeki 2000).
2. The number of Least Developed Countries (LDC) has risen from 25 to 49 in the past 30 years.
3. Emmanuel Wallerstein attempted to broaden the scope of dependency theory so as to be able to explain the emergence of countries such as South Korea and Taiwan. His theory of international relations, based on the dependency paradigm, is now commonly described as World System Theory. Wallerstein argued that it is important to focus on the world system as a whole. He proposed that a capitalist world economy has been in existence since the sixteenth century. Like Frank, he identified a core and periphery within the global market, but included a further category, the semi-periphery. He maintained that the core economies were built on industry and differentiated agriculture whereas the periphery is tied to monoculture. The semi-periphery occupies a position roughly in between the two. For Wallerstein, the countries occupying this sector are important because they help to keep the system in check. The semi-periphery is proof that progress can be made and, additionally, these countries ensure that no united front is likely to be formed against the core. The concept of a semi-periphery allowed Wallerstein to explain developments such as the rise of the 'Asian Tigers' in a manner that would have proved difficult within conventional dependency theory analysis (Randall and Theobald 1998, pp. 145–46).
4. See B. Stallings, 'International Influence on Economic Policy: Debt, Stabilisation and Structural Reform' in S. Haggard and R. Kaufman (eds), *The Politics of Economic Adjustment* (Princeton: Princeton University Press, 1992). While anxious to avoid overt reductivism, Stallings argues that international influences are to blame for the tide of liberalisation that swept Africa from the late 1980s onwards.
5. The term was coined by John Williamson in 1990 and proscribed macroeconomic reforms together with more effective financial management. Williamson had not intended for the term or his analysis to become synonymous with development initiatives in the South. He listed ten policy reforms that he claimed were generally agreed upon in Washington that, if enacted, would be to the benefit of the majority of Latin American countries. The ten reforms are fiscal discipline, public expenditure priorities, tax reform, financial liberalisation, exchange rates, trade liberalisation, foreign direct investment, privatisation, deregulation and property rights. The phrase rapidly entered the development parlance and has been used as a blanket term to describe the policies advocated by major aid donors and international

financial institutions (IFIs). Williamson argues in a lecture delivered in 2004 that the consensus, in the sense that he outlined it, has 'evaporated', largely because of a split between the Bush White House and the major IFIs on issues such as fiscal policy, income distribution and capital account convertibility (Williamson 2004).
6. Hirst and Thompson cite K. Ohmae and R. B. Reich as proponents of this view (Hirst and Thompson 1999, pp. 262, 270).
7. The G-20 and G-90 coalitions evolved in the WTO in response to the 2003 Cancun Agenda. The G-20 includes Brazil, India, China, South Africa, Egypt, Pakistan and Nigeria. The G-90 comprises countries drawn from the ACP, the African Group, the LDC Group and the Small-Island State Group.
8. Western Europe (the EU, together with Switzerland, Turkey, Norway, Malta, Liechtenstein and the states of the former Yugoslavia) saw 18 per cent of combined GDP being traded amongst the bloc (WTO figures for 2000) as opposed to slightly less than 11 per cent between East Asian countries and North America. World Trade Organisation, *Annual Report 2000, Volume II, International Trade Statistics*, derived from tables cited in Hirst and Thompson (2002, pp. 258–59).

4 The 'Everything But Arms' Initiative

1. Council Regulation (EC) No 416/2001 of 28 February 2001 amending Regulation 2820/98 applying to a multiannual scheme of generalised tariff preferences for the period from 1 July 1999 to 31 December 2001 so as to extend duty-free access without any quantitative restrictions to products originating in the LDCs, *OJ L60/43*.
2. See Council Regulation (EC) No 416/2001 (28 February 2001).
3. See Council Regulation (EC) No 1602/2000 of 24 July 2000 amending Regulation (EEC) No 2454/93 laying down provisions for the implementation of Council Regulation (EEC) No 2913/92 establishing the Community Customs Code, *OJ L 188 of 26 July 2000*.
4. In April 2005 the WTO Appellate Body ruled in favour of a complaint brought by Brazil, Thailand and Australia against the EU sugar regime, especially with regards to export subsidies.

5 Trade Policy, Poverty Alleviation and Sustainable Development

1. In terms of food security, on a national level, self-sufficiency entails 'meeting food needs as far as possible from domestic supplies and minimising dependence on food trade' whilst the concept of self-reliance 'takes into account the possibilities offered by international trade' (UNESCAP Bulletin on Asia-Pacific Perspectives 2004–2005).
2. Zero-growth advocates such as Herman Daly (1992, 1996) argue that bolstering economic growth will only serve to hasten global environmental degradation.
3. In February 2001, the Council adopted the so-called 'EBA (Everything But Arms) Regulation' (Regulation (EC) 416/2001), granting duty-free access to imports of all products from LDCs without any quantitative restrictions,

except with regard to arms and munitions. At present, 50 developing countries belong to the category of LDCs. Only imports of fresh bananas, rice and sugar were not fully liberalised immediately. Duties on these products were to be gradually reduced until duty-free access was achieved (restrictions on bananas were removed in January 2006; restrictions will be removed for sugar by July 2009 and for rice by September 2009). The EBA Regulation foresees that the special arrangements for LDCs should be maintained for an unlimited period of time and not be subject to the periodic renewal of the Community's scheme of generalised preferences (see Chapter 4).

4. That this position is largely accepted as a truism is acknowledged by the EU Trade Commissioner, Peter Mandelson (2005), in a speech in Edinburgh in July 2005. He argues that the public should not fall into 'the simplistic trap of believing that the abolition of all or part of the CAP is the solution to the problems of Africa.'
5. Botswana is a good case in point. The country represents an economic success story in a region of often failing or faltering economies. It managed to achieve and sustain growth rates of over 7 per cent between 1985 and 1999, accruing nearly $6 billion in foreign exchange reserves in the process, as well as diversifying its economy away from the mining sector. Despite this, 47 per cent of the population remains below the poverty line, confined, in many instances, to marginal, environmentally fragile areas (Leith *et al.* 1999, pp. 1–21). The United Nations Development Programme's (2006, p. 285) Human Development Index (HDI) ranks Botswana a lowly 131 out of 177 countries.
6. 'Farm-to-fork' legislation, based on the White Paper, was passed in January 2002 (see European Commission 2000a, p. 24). The European Food Safety Authority (EFSA) was established to oversee the implementation of the policy. See Regulation (EC) No 178/2002 of the European Parliament and of the Council of 28 January 2002 laying down the general principles and requirements of food law, establishing the EFSA and laying down procedures in matters of food safety.
7. Recent research suggests that the future of the banana may be threatened by the black sigatoka (*mycosphaerella fijiensis*) fungus which attacks the ubiquitous Cavendish variety. As nearly all plantation-farmed bananas are of this variety, a rapid spread of the fungus could be catastrophic for the industry (*New Scientist* 13 May 2006).
8. Traditionally, coffee is a shade-grown crop that requires little fertilizer in order to develop. However, to increase yields, 'full-sun' variants were developed. These require substantial agrochemical input in order to propagate successfully (Clay 2004).
9. The EU will liberalise around 95 per cent of its imports from South Africa within 10 years, whilst the respective figures on the South African side are around 86 per cent in 12 years [*OJL 311 of 4 December 1999*].

6 Agriculture, Food Security and Rural Development

1. The Caribbean ACP members, whilst traditionally agricultural in focus, have increasingly moved towards the service sector and manufacturing.

Services now account for the greatest percentage of Caribbean GDP and provide approximately 50 per cent of all employment in the region (Boudeville 2004, p. 5).
2. Data from the majority of Pacific states is sorely lacking in places. Despite the fact that agriculture accounts for nearly 50 per cent (and above) of employment in the Cook Islands, Fiji, Samoa, the Solomon Islands, Tonga and Vanuatu, poverty levels are considerably lower (with the exception of Vanuatu) than those in many African countries (Sharma 2006, p. 16).
3. The EU itself is somewhat reticent about publicising the exact percentage of its budget allocated to the CAP. In a booklet intended for general consumption, the Commission states only that this figure has dropped to 'less than 50 percent of the total EU budget'. The cost of the Common Agricultural Policy (CAP) is, the publication suggests, 'hardly a high price to pay for a healthy supply of food and a living countryside' (European Commission 2004a, p. 28).
4. To place this figure in perspective, the Mozambiquan GDP stood at $4.3 billion in 2003 (World Bank 2005a).
5. Recent budgetary figures (2005) indicate that ACP countries received €111.05 million in support for their agricultural sectors out of a total budget of €2723.15 million, together with €98.82 million for food security and a further €9.09 million for general environmental projects. Funding of €9 million was also made available for forestry by the European Commission. These figures do not include funding for rural development programmes which often form part of other budgetary lines, such as infrastructure or institutional capacity building (European Commission 2005a, pp. 154–5, Table 7.12).

7 Energy Provision, Poverty Alleviation and Sustainable Development

1. In an effort to wean its electricity production off diesel generators, Fiji has begun to concentrate on small-scale hydroelectric projects on the main island of Viti Levu.

9 EU-ACP Economic Partnership Agreements

1. During the 25-year period covered by the Lomé, ACP countries actually saw their share of the EU market fall, from 7 per cent in 1975 to 2.8 per cent in 1995 (Dickson 2004, p. 46).
2. The terms 'substantially all' and 'reasonable period of time' are not quantified by the WTO. The EU has decided that 90 per cent represents 'substantially all' and that 12 years represent a 'reasonable period of time'.
3. See W. Sandholtz and J. Zysman (1994, pp. 189–209). Their approach can be said to be a mixture of neofunctionalism, theories of domestic politics and a focus on changes in the international structure, that is, the economic decline of the USA. See also R. Keohane and S. Hoffman (2003,

pp. 237–55). The authors argue that a combination of supranationalism and intergovernmentalism is required in order to explain the *relance*. They assert that neofunctional elements such as spillover together with an assessment of the international political economy, as well as preference-convergence, must all be taken into account when attempting to understand the processes that shaped the Single European Act (SEA). See also W. Sandholtz (1994). Sandholtz combines Neofunctionalist elements, domestic politics, interstate politics and intergovernmentalism in an attempt to offer an explanation for member states choosing Monetary Union as the preferred method of governing monetary policy.

4. For instance, the liberal intergovernmentalism espoused by Hoffman and Keohane, the Neofunctionalist-transactional model (Sandholtz 2003) and the multilevel governance paradigm (Hooghe and Marks 2001).
5. For example, see A. Malamud (2001, pp. 1–22).
6. On 6 July 1965, the French Government announced the recall of its representative to the EU, stating that it was no longer prepared to participate in the meetings of the Council of Ministers. France's chair on the Council thus became 'empty'. De Gaulle objected to efforts by the Commission to introduce majority voting in the Council, reform of the CAP, and the allocation of additional budgetary powers to the European Parliament. In January 1966, six months after the French precipitated the 'Empty-Chair Crisis', an informal agreement was reached by the six member states in an attempt to resolve the impasse. The resultant Luxembourg Compromise effectively enshrined the principle of unanimity in Council meetings despite there being no legal basis for doing so. The Compromise successfully bestowed veto powers on each of the six, severely curtailing the power of the Commission.
7. The dispute was sparked by a biscuit war in 2005. Vanuatu imposed tariffs and quotas on Fijian biscuits, ostensibly to stimulate domestic production. Fiji responded by imposing tariffs on kava from Vanuatu and threatening further measures, including a full cessation of trading relations.
8. In February 2001, the EU adopted the so-called 'EBA (Everything But Arms) Regulation' (Regulation (EC) 416/2001), granting duty-free access to all imports from LDCs without any quantitative restrictions, with the exception of arms and munitions. Only imports of fresh bananas, sugar and rice were not fully liberalised immediately (see Chapter 4 for a full discussion on the initiative).

Bibliography

Abramovitz, M. (1993) 'Catching up, Forging Ahead and Falling behind' in Seligson, M. A. and Passé-Smith, J. T. (eds) *Development and Underdevelopment: The Political Economy of Inequality* (Boulder/London: Lynne Rienner).

ACP-EC Energy Facility (2006) Newsletter, May.

ACP-EU (2001) (Joint) Parliamentary Assembly, ACP-EU 3171/01/fin Resolution, Libreville (Gabon) 19–22 March.

ACP-EU Courier (2000) September.

ACP-EU Courier (2001) May–June.

ACP-EU Courier (2002) September–October.

ACP Group (2004) 'Developmental Aspects of Regional Trade Agreements and Special and Differential Treatment in WTO Rules: GATT 1994 ARTICLE XXIV and the Enabling Clause', Communication by the Mission of Botswana on behalf of the ACP Group of States, TN/RL//W/155, 28 April.

Action for Southern Africa (ACTSA) (2000) *Trade and Development Update* Volume 1, Issue 2, September.

Action for Southern Africa (ACTSA) (2001a) 'Freedom to Grow: Bringing Down the Barriers to Southern Africa's Trade with Europe', Trade Briefing Paper, December.

Action for Southern Africa (ACTSA) (2001b) *Trade and Development Update* Volume 2, Issue 2, December.

Action for Southern Africa (ACTSA) (2002a) *Trade and Development Update*, Volume 1, Issue 2, September.

Action for Southern Africa (ACTSA) (2002b) *Trade and Development Update*, Volume 2, Issue 5, November.

Action for Southern Africa (ACTSA) (2002c) 'With Friends Like These ...', Briefing Paper, October.

Action for Southern Africa (ACTSA) (2003) *Trade and Development Update*, Volume 3, Issue 2, November.

African Ministerial Conference on Water (AMCOW) – Technical Advisory Committee (TAC) and the EU (2003) 'The EU Water Initiative: Outline Strategy and 2004–2005 Work Programme', Working Document as Discussed at the Pan-African Implementation and Partnership Conference on Water, December.

African Water Issues Research Unit (University of Pretoria) (2000) 'Hydropolitical Hotspots in Southern Africa: Will There be a Water War?' African Dialogue Lecture Series, 24 February.

Allen, T. (2000) 'EU Trade with ACP Countries', *Statistics in Focus: External Trade*, Theme 6, Eurostat Catalogue Number KS-NO-02-003-EN-N (EU).

Amin, S. (1972) 'Underdevelopment and Dependence in Black Africa: Origins and Contemporary Forms', *Journal of Modern African History*, Volume 10, Issue 4.

Annecke, W. (2001) 'Impact of Women's Energy Group on National Policy: South Africa' in Misana, S. and Karlsson, G. V. (eds) *Generating Opportunities: Case Studies on Energy and Women* (New York: UNDP).

Ashton, P. (2000) 'Southern Africa Water Conflicts: Are They Inevitable or Are They Preventable?' in Solomon, H. and Turton, A. (eds) *Water Wars: Enduring Myth or Impending Reality?* African Dialogue Monograph Series 2 (Durban: ACCESS).

Association Des Organisations Professionnelles Du Commerce Des Sucre Our LesPays De l'Union Europeenne (ASSUC) (2001) 'EBA: An Impact Assessment for the Sugar Sector', Impact Study, 25 January.

Axford, B. (1995) *The Global System: Economics, Politics and Culture* (Cambridge: Polity Press).

Ayres, R. U. (1995) 'Economic growth: Politically necessary but not environmentally friendly', *Ecological Economics*, Volume 15, Number 2.

Bailey, M. (2002) 'Europe's Double Standards: How the EU Should Reform Its Trade Policies with the Developing World', Oxfam Briefing Paper Number 22.

Baran, P. (1957) *The Political Economy of Growth* (New York: Monthly Review Press).

Baran, P. (1980) 'On the Political Economy of Backwardness' in Agarwala, A. N. and Singh, S. P. (eds) *The Economics of Underdevelopment* (Oxford/New York: Oxford University Press).

Barry, J. (1999) *Rethinking Green Economics* (London: Sage).

Basu, A. and Srinivasan, K. (2002) 'Foreign Direct Investment in Africa: Some Case Studies', IMF Working Paper, WP/02/61 March.

Bates, R. H. (1993) 'Governments and Agricultural Markets in Africa' in Seligson, M. A. and Passé-Smith, J. T. (eds) *Development and Underdevelopment: The Political Economy of Inequality* (Boulder/ London: Lynne Rienner).

Bayart, J -F. (1991) 'Finishing with the Idea of the Third World: The Concept of the Political Trajectory' in Manor, J. (ed.) *Rethinking Third World Politics* (Harlow: Longman).

Bayliss, K. (2002) 'Water Privatization in SSA: Progress, Problems and Policy Implications', Conference Paper, Development Studies Association Annual Conference, University of Greenwich, 9 November.

Bhagwati, J. and Meyer, A. (2002) 'Special Report: The Poor's Best Hope: Trading for Development: By Invitation', *The Economist,* Volume 363, 22 June.

Bierman, F. (2000) 'Rising Tide of Green Unilateralism in World Trade Law: Options for Reconciling the Emerging North–South Conflict', PIK Report Number 66.

Bilal, S. (2002a) 'The Compatibility of EPAs with WTO Rules: A Moving Target?' *Trade Negotiation Insights*, Volume 1, Number 4, December.

Bilal, S. (2002b) 'Who Will Negotiate with the EU? In Search of an ACP-EU Negotiating Framework', *Trade Negotiation Insights,* Volume 1, Number 1, February.

Bilal, S. and Rampa, F. (2006) 'Alternatives (to) EPAs: Possible Scenarios for the Future ACP Trade Relations with the EU', ECDPM Policy Management Report 11, February.

Bilal, S., de Burgh, M., Rampa, F. and Van Hove, K. (2006a) 'EPA Development Support: Consequences of GAERC Conclusions for EPAs', ECDPM *InBrief* 16A, November.

Bilal, S., Donnat, C., Gerard, N., Rampa, F. and Van Hove, K. (2006b) 'Overview of the Regional EPA Negotiations' *InBrief* 14B, November.

Bird, A., Soligna-Lacomte, H. B. and Wilson, T. (2001) 'Evaluation of EC Country Strategy: Namibia 1996–2000', European Commission, EuropeAid, April.

Blundell, J. (1995) 'Foreword' in Ridley, M. (1995) *Down to Earth: A Contrarian View of Environmental Problems* (London: Institute of Economic Affairs).

Bomberg, E. (1998) *Green Parties and Politics in the European Union* (London/New York: Routledge).

Bornschier, V. and Chase-Dunn, C. (1993) 'Transnational Penetration and Economic Growth' in Seligson, M. A. and Passé-Smith, J. (eds) *Development and Underdevelopment: The Political Economy of Inequality* (Boulder/London: Lynne Rienner).

Borrell, B. and Hubbard, L. (2000) 'Global Economic Effects of the EU Common Agricultural Policy', *Economic Affairs*, Volume 20, Issue 2, June.

Boudeville, N. (2004) 'Sustainability Impact Assessment (SIA) of the EU-ACP Economic Partnership Agreements', Report for the European Commission, February.

Brenton, P. and Manchin, M. (2002) 'Making EU Trade Agreements Work: The Role of Rules of Origin', Centre for European Studies Working Document Number 183.

Broca, S., Stamoulis, K. and Zezza, A. (2003) 'Agriculture in Poverty Alleviation and Economic Development' in Bruinsma, J. (ed.) *World Agriculture: Towards 2015/2030: An FAO Perspective* (London: Earthscan).

Brohman, J. (1997) *Popular Development: Rethinking the Theory and Practice of Development* (Oxford: Blackwell).

Brown, W. (2004) 'From Uniqueness to Uniformity? An Assessment of EU Development Aid Policies' in Arts, K. and Dickson, A. (eds) *EU Development Cooperation* (Manchester: Manchester University Press).

Brüntrup, M. (2006) 'Everything But Arms (EBA) and the EU-Sugar Market Reform – Development Gift or Trojan Horse?' German Development Institute Discussion Paper , October.

Bucknall, J. (1999) 'Poverty and Environment', *Annual Review of Environmental Matters at the World Bank*, Report Number 20104, December.

Bulmer, S. (2003) 'Domestic Politics and European Community Policy Making' in Nelsen, B. F. and Stubb, A. C -G. (eds) *The European Union: Readings on the Theory and Practice of European Integration*, 3rd edn (London/Boulder: Lynne Rienner).

Burke, T. (2001) 'Ten Pinches of Salt: A Reply to Bjørn Lomborg', Green Alliance Paper, August.

The Business Times (17 June 2004).

Byers, A. (2002) 'The Lesotho Highlands Water Project: Supporters, Critics, and Mountain Voices', The Mountain Institute – Case Study on Mountain Infrastructure: Access Communication and Energy. Consultation for the UNEP/Bishkek Global Mountain Summit.

Cadot, O., de Melo, J. and Olarreaga, M. (1999) 'Asymmetric Regionalism in Sub-Saharan Africa: Where do we stand?' Annual World Bank Conference on Development Economics in Europe, Paris, 21–23 June.

Cardoso, F. H. and Faletto, E. (1979) *Dependency and Development in Latin America* (Berkeley: University of California Press).

Carruthers, J. (1995) *The Kruger National Park: A Social and Political History* (Pietermaritzburg: University of Natal Press).

Castro, J. (1972) 'Environment and Development: The Case of Developing Countries', *International Organisation*, Volume 26, Number 2, Spring.

Cernat, L., Laird, S., Monge-Roffarello, L. and Turrini, A. (2003) 'The EU's Everything But Arms Initiative and the Least Developed Countries', United Nations University/World Institute for Development Economics Research, Discussion Paper Number 2003/47, June.

Chabal, P. (1997) 'Apocalypse Now? A Postcolonial Journey into Africa', Inaugural Lecture, King's College London, 12 March.

Chabal, P. and Daloz, J. P. (1999) *Africa Works: Disorder as Political Instrument* (Oxford: James Currey).

Chisholm, N. (2001) 'Poverty-Environment-Gender Linkages', *Development Assistance Committee Journal,* Volume 2, Number 4.

Chryssochoou, D. N. (1998) 'Democracy and Integration Theory in the 1990s: A Study in European Polity-Formation', Jean Monnet Working Paper Number 14, University of Catania, June.

Chryssochoou, D. N. (2001) *Theorising European Integration* (London: Sage).

Clapp, J. (2000) 'The Global Economy and Environmental Change in Africa' in Stubbs, R. and Underhill, G. R. D. (eds) *Political Economy and the Changing Global Order,* 2nd edn (Canada: Oxford University Press).

Clay, J. (2004) *World Agriculture and the Environment* (Washington DC: Island Press).

Cock, J. and Fig, D. (2001) 'The Impact of Globalisation on Environmental Politics in South Africa, 1990–2002', *African Sociological Review,* Volume 5, Issue 2.

Conca, K. and Dabelko, G. D. (2004) *Green Planet Blues: Environmental Politics from Stockholm to Johannesburg,* 3rd edn (Boulder/Oxford: Westview Press).

Cornia, G. A., Jolly, R. and Stewart, F. (1987) *Adjustment with a Human Face: Protecting the Vulnerable and Promoting Growth* (Oxford: Oxford University Press).

Corporate Europe Observatory (CEO) (2003) 'Evian: Corporate Welfare or Water for All?' *InBrief* 6, May.

Council of Canadians' Blue Planet Project and the Polaris Institute (2002) 'Facts on the World's Top Water Companies', Fact Sheet 1.

Cram, L. (1997) *Policy Making in the EU: Conceptual Lenses and the Integration Process* (New York: Routledge).

Crichton, M. (2004) *State of Fear* (New York: HarperCollins).

Cromwell, D. (2001) *Private Planet: Corporate Plunder and the Fight Back* (Charlbury: Jon Carpenter).

Cuervo, J. and Gandhi, V. P. (1998) 'Carbon Taxes: Their Macroeconomic Effects and Prospects for Global Adoption – A Survey of the Literature', IMF Working Paper, WP/98/73, May.

Daly, H. E. (1992) *Steady-State Economics,* 2nd edn (London: Earthscan).

Daly, H. E. (1993) 'Sustainable Growth: An Impossible Theorem' in Daly, H. E. and Townsend, K. N. (eds) *Valuing the Earth: Economics, Ecology and Ethics* (Cambridge, Massachusetts: MIT Press).

Daly, H. E. (1996) *Beyond Growth: The Economics of Sustainable Development* (Boston: Beacon Press).

Danish Government Committee on Scientific Dishonesty (2002) 'Annual Report', <http://www.forsk.dk/uvvu/publ/uvvu2002.pdf>

Danish Presidency of the EU (2002) 'The European Union Energy Initiative for Poverty Eradication and Sustainable Development: A Partnership Initiative', Press Release, January.

Danish Research Institute of Food Economics (2003) 'WTO Negotiations and Changes in Agricultural and Trade Policies: Consequences for Developing Countries', Policy Brief, 8 June.

Dasgupta, S., Mody, A., Roy, S. and Wheeler, S. (1995) 'Environmental Regulation and Development: A Cross-Country Empirical Analysis', World Bank Policy Research Department Working Paper Number 1448, April.

Davidson, O., Tyani, L. and Afrane-Okesse, Y. (2002) 'Climate Change, Sustainable Development and Energy: Future Perspectives for South Africa', Energy and Development Research Centre, University of Cape Town, Prepared for the OECD.

De Waal, D. (2003) 'Prospects for the Poor: Water Reforms and Private Sector Participation in Dar es Salaam, Tanzania', Water Aid Paper.

Dearden, S. J. H. (2002) 'Does the European Union's Development Policy Have Any Future?' European Development Policy Study Group, Discussion Paper Number 24, Manchester Metropolitan University, September.

Department for International Development (UK) (2002) 'Energy for the Poor: Underpinning the Millennium Development Goals', August.

Dickson, A. (2004) 'The Unimportance of Trade Preferences' in Arts, K. and Dickson, A. (eds) *EU Development Cooperation* (Manchester: Manchester University Press).

Dieter, H. and Melber, H. (2000) 'No Future for SADC? Perspectives for Regional Integration in Southern Africa after the Mauritius Summit', Institut für Entwicklung und Frieden Report, Heft 43/2000.

Dikotla, M. and Verhoef, G. (2002) 'The Development of Trade in SADC: The First Decade of Trade in the New Regional Formation, 1980–1989', Conference Paper, International Economic History Congress, Buenos Aires, 23 July.

Dobson, A. (1991) 'The Special Relationship and European Integration', *Diplomacy and Statecraft*, Volume 2, Number 1, March.

Dobson, A. (1995) *Green Political Thought*, 2nd edn (London/New York: Routledge).

Doherty, M. (2006) 'Negotiating Economic Partnership Agreements: Sanitary and Phytosanitary Measures', ECDPM *InBrief* 13A, 7 July.

Dollar, D. and Collier, P. (2001) *Globalisation, Growth and Poverty: Building an Inclusive World Economy* (Oxford: Oxford University Press).

Dos Santos, T. (1993) 'The Structure of Dependence' in Seligson M. A. and Passé-Smith, J. T. (eds) *Development and Underdevelopment: The Political Economy of Inequality* (Boulder/London: Lynne Rienner).

Duff, A. (ed.) (1997) *The Treaty of Amsterdam, Text and Commentary* (London: Sweet and Maxwell).

Dumont, R. (1988) *False Start in Africa* (Translated by Ott, P. O.) (London: Earthscan).

Esty, D. and Levy, M. (2006) *Pilot 2006: Environmental Performance Index* (New Haven: Yale Centre for Environmental Law and Policy).

European Centre for Development Policy Management (ECDPM) (2002) 'Implementing EC Development Policy: Debates and Emerging Approaches to Poverty Reduction', Maastricht, August 2002.

European Centre for Development Policy Management (ECDPM) (2006a) 'Update on Regional EPA Negotiations: Caribbean – EU Economic Partnership Agreement', *ECDPM InBrief 15C*, December.

European Centre for Development Policy Management (ECDPM) (2006b) 'Update on Regional EPA Negotiations: Central Africa – EU Economic Partnership Agreement', *ECDPM InBrief 15A,* November.

European Centre for Development Policy Management (ECDPM) (2006c) 'Update on Regional EPA Negotiations: Pacific – EU Economic Partnership Agreement', *ECDPM InBrief 15D*, November.

European Centre for Development Policy Management (ECDPM) (2006d) 'Update on Regional EPA Negotiations: ESA – EU Economic Partnership Agreement', *ECDPM InBrief 15E*, November.

European Centre for Development Policy Management (ECDPM) (2006e) 'Update on Regional EPA Negotiations: SADC – EU Economic Partnership Agreement', *ECDPM InBrief 15F* December.

European Commission (1996) *Green Paper on Relations between the European Union and the ACP Countries on the Eve of the 21st Century. Challenges and Options for a New Partnership*, COM(96)570 final, Brussels, 20 November.

European Commission (1999) *Assessing the Integrated Regional Potential for theWidespread Introduction and Application of Appropriate Renewable Energy Technologies within SADC*, Final Report, September.

European Commission (2000a) 'European Policy to Support Rural Development', Policy Orientation Paper, February.

European Commission (2000b) 'Everything But Arms Proposal: Possible Impacts on the Agriculture Sector', Impact Assessment, November.

European Commission (2000c) 'Integration of Food Security Objectives within a Poverty Reduction Framework', Concept Paper, February.

European Commission (2000d) *White Paper on Food Safety*, COM(1999) 719 final, Brussels, 12 January.

European Commission (2001a) *A Sustainable Europe for a Better World: A European Union Strategy for Sustainable Development*, Communication from the Commission, COM(2001)264final, Brussels, 15 May.

European Commission (2001b) 'EU Approves "Everything But Arms" Initiative', Ref: EC01-015EN, 26 February.

European Commission (2001c) *On the Sixth Environmental Action Programme of the European Community, 'Environment 2010: Our Future – Our Choice'*. Communication from the Commission to the Council, the European Parliament, the Economic and Social Committee and the Committee of the Regions, COM(2001)31 final, Brussels, 24 January.

European Commission (2001d) *Synergy Programme Projects Selected in 2001*, Summary.

European Commission (2002a) *Annual Report 2001 from the Commission to the Council and the European Parliament on the EC Development Policy and the Implementation of the External Assistance*, COM(2002)490 final, Brussels, 12 September.

European Commission (2002b) *Annual Report 2001 on the EC Development Policy and the Implementation of the External Assistance* (Brussels: EuropeAid Cooperation Office).

European Commission (2002c) *Communication from the Commission to the Council and the European Parliament. Mid-Term Review of the Common Agricultural Policy*, COM(2002)394 final, Brussels, 10 July.

European Commission (2003a) *Annex to the Communication from the Commission to the Council and the European Parliament. The World Summit on Sustainable Development One Year On: Implementing Our Commitments*, Commission Staff Working Paper SEC(2003)1471, Brussels, 23 December.

European Commission (2003b) *Making Trade Work for Development: Putting Theory into Practice – Case Reports of European Union Development Projects around the World* (Brussels: EuropeAid Cooperation Office).
European Commission (2003c) *Reform of the Common Agricultural Policy: Medium-Term Prospects for Agricultural Markets and Income in the European Union, 2003–2010,* Brussels, December.
European Commission (2003d) *Reforming the European Union's Sugar Sector: Summary of Impact Assessment Work,* Commission Staff Working Document SEC(2003)1022, Brussels, 23 September.
European Commission (2003e) *The European Union Energy Initiative for Poverty Eradication and Sustainable Development,* Brussels, September.
European Commission (2004a) *Communication from the Commission to the Council and the European Parliament. Communication on the Future Development of the EU Water Initiative and the Modalities for the Establishment of a Water Facility for ACP Countries,* COM(2004)43 final, Brussels, 26 January.
European Commission (2004b) *The Common Agricultural Policy Explained* (Eugène Leguen de Lacroix: Germany)
European Commission (2004c) *The Future of the EU Energy Initiative and the Modalities for the Establishment of an Energy Facility for ACP Countries,* Communication from the Commission to the Council and the European Parliament, COM(2004)711final, Brussels, 26 October.
European Commission (2005a) *Annual Report 2005 on the European Community's Development Policy and the Implementation of External Assistance in 2004* (Brussels: EuropeAid Cooperation Office).
European Commission (2005b) *Reforming the European Union's Sugar Sector: Update of Impact Assessment,* Commission Staff Working Document SEC(2005)808, Brussels, 22 June.
European Commission (2006a) '"EBA" – Everything But Arms Initiative: User's Guide to the EU GSP's – Special Arrangements for Least Developed Countries', *Trade Issues,* January (updated).
European Commission (2006b) 'European Union to Commit €22 Bbillion in Aid to Africa, the Caribbean and the Pacific Islands', Press Release, Reference IP/06/710, 1 June.
European Commission (2006c) *Global Europe: Competing in the World – A Contribution to the EU's Growth and Jobs Strategy,* COM(2006)567, Brussels, 4 October.
European Commission (2006d) *The Budget of the European Union for 2006.*
European Commission Budget (2007) *The Budget of the European Union for 2007.*
European Commission (2007b) <http://ec.europa.eu/unitedkingdom/information/policy_briefs/bb01a_en.htm>
European Council (2000) 'Lisbon European Council – Presidency Conclusions', Brussels, 23–24 March.
European Council (2001) 'Göteborg European Council – Presidency Conclusions', Brussels, 15–16 June.
European Council (2005) 'Brussels European Council – Presidency Conclusions', Brussels, 16–17 June.
European Council Directive (1967) 67/548/EEC of 27 June on the approximation of laws, regulations and administrative provisions relating to the classification,

packaging and labelling of dangerous substances, *Official Journal 196, 16 August 1967, P. 0001 – 0098 English Special Edition: Series I Chapter 1967 P. 0234.*

European Council Regulation (2000) No 1602/2000 of 24 July amending Regulation (EEC) No 2454/93 laying down provisions for the implementation of Council Regulation (EEC) No 2913/92 establishing the Community Customs Code, *OJ L 188 of 26 July 2000.*

European Council Regulation (2001) No 416/2001 of 28 February amending Regulation 2820/98 applying to a multiannual scheme of generalised tariff preferences for the period from 1 July 1999 to 31 December 2001 so as to extend duty-free access without any quantitative restrictions to products originating in the Least Developed Countries, *OJ L60/43.*

European Environment Agency (EEA) (2002) 'Energy and Environment in the European Union', Environmental Issue Report, Number 31, Executive Summary.

European Environment Agency (EEA) (2004a) 'Annual European Community Greenhouse Gas Inventory 1990–2002 and Inventory Report 2004', Technical Report, Number 2/2004, 14 July.

European Environment Agency (EEA) (2004b) 'EU15 Greenhouse Gas Emissions Decline after Two Years of Increases', Press Release, 15 July.

Energy Information Administration (EIA) (2002) 'Southern African Power Pool', Analysis Brief, November.

Energy Information Administration (EIA) (2006) 'Southern Africa (SADC)', Regional Analysis Brief, June.

Energy Information Administration (EIA) (2007) 'South Africa', Country Analysis Brief, April.

European Union at the United Nations, 'EU Humanitarian Aid for Crisis in Southern Africa', Press Release, Reference EC02-212EN, 18 October 2002, <http://europa-eu-un.org/articles/hu/article_1694_hu.htm>

European Union at the United Nations (2003) 'Water for Life: Europe Day at the World Water Forum', Press Release, Reference: EC03-074EN, 21 March.

European Union (Delegation in Tanzania) (2003) *Newsletter,* October <http://www.deltza.cec.eu.int/en/newsletter_oct_2003/content.htm>.

European Union Water Initiative (EUWI) (2002) '"Water for Life": European Union Launches Global Initiative to Achieve World Summit Goals', Press Release, IP/02/1265, Johannesburg, 3 August.

Farinelli, U. (ed.) (1999) *Energy as a Tool for Sustainable Development for African, Caribbean and Pacific Countries* (European Commission/UNDP).

Frank, A. G. (1967) *Capitalism and Underdevelopment in Latin America: Historical Studies of Chile and Brazil* (New York/London: Monthly Review Press).

Frank, A. G. (1996) 'The Development of Underdevelopment' in Jameson, K..P. and Wilber, C. K. (eds) *The Political Economy of Development and Underdevelopment,* 6th edn (New York: McGraw-Hill).

Fischler, F. (2003a) Participation in the 7th Churchill Conference, Zurich, Speech/03/434, 26 September.

Fischler, F. (2003b) 'The New, Reformed Agricultural Policy', Final press conference after the decision at the Council on agriculture, Luxembourg, Speech/03/326, 26 June.

Food and Agriculture Organisation (2006) *Food Outlook – Global Market Analysis,* Number 2, December.

George, C. and Kirkpatrick, C. (2003) 'Sustainability Impact Assessment of Proposed WTO Negotiations: Preliminary Overview of Potential Impacts of the Doha Agenda – Assessment of Individual Trade Measures', Institute for Development Policy and Management, University of Manchester, 4 June.

Georgieva, K. (Chair) (2002), *Making Sustainable Commitments: An Environment Strategy for the World Bank* (Washington DC: World Bank).

Gereffi, G. (1997) 'The Elusive Last Lap in the Quest for Developed Country Status' in Mittelman, J. H. (ed.) *Globalization: Critical Reflections* (Boulder/London: Lynne Rienner).

Gibb, R. (2002) 'Developments in the European Union: Implications for Southern Africa' in *SADC-EU Relations: Looking Back and Moving Ahead* (Braamfontein: SA Institute of International Affairs).

Godfrey, C. (2002) 'Stop the Dumping! How EU Agricultural Subsidies are Damaging Livelihoods in the Developing World', Oxfam Briefing Paper Number 31.

Goldstein, J. and Keohane, R. O. (1993) 'Ideas and Foreign Policy: An Analytical Framework in Goldstein, J. and Keohane, R. O. (eds) *Ideas and Foreign Policy: Beliefs, Institutions and Political Change* (Ithaca: Cornell University Press).

Goodison, P. and Stoneman, C. (2005) 'Trade, Development and Cooperation: Is the EU helping Africa?' Melber, H. (ed.) *Trade, Development, Cooperation: What Future for Africa?* (Uppsala: Nordiska Afrikainstitutet).

Green, D. and Griffith, M. (2002) 'The Rough Guide to the CAP', CAFOD Policy Paper.

Greenpeace and the Intermediate Technology Development Group (2002)'Sustainable Energy for Poverty Reduction: An Action Plan', Joint Report, 30 August.

Griffith, M. and Powell, S. (2007) 'Partnership under Pressure: An Assessment of the European Commission's Conduct in the EPA Negotiations', Tearfund/Tradecraft.

Griswold, D. T. (2001) 'Trade, Labour, and the Environment: How Blue and Green Sanctions Threaten Higher Standards', Centre for Trade Policy Studies Study Paper (CATO Institute), Washington DC, Number 15, August.

Groom, A. J. R. (2003) 'Neofunctionalism: A Case of Mistaken Identity' in Nelsen, B. F. and Stubb, A. C -G. (eds) *The European Union: Readings on the Theory and Practice of European Integration*, 3rd edn (London/Boulder: Lynne Rienner).

Grossman, G. and Krueger, A. (1994) 'Economic Growth and the Environment', NBER, Working Paper No. 4634.

The Guardian (9 January 2003).

Gunessee, S. (2002) 'Major Issues of Contention arising from Current EU and ACP Negotiating Positions', Conference Report, The Cotonou Trade Negotiations: Building European Solidarity for Southern Africa, ACP House, Brussels, 10 October.

Hall, K. O. and Blake, B. W. (1979) 'The Emergence of the African, Caribbean, and Pacific Group of States: An Aspect of African and Caribbean International Cooperation', *African Studies Review*, Volume XXII, Number 2, September.

Hakonsen, L. and Mathiesen, L. (1997) 'Carbon Dioxide Stabilisation may be a "No-Regrets" Policy', *Environmental and Resource Economics*, Volume 9.

Hamburger, F. (2002) 'EU Initiative on Energy for Poverty Eradication and Sustainable Development', Speech to the Global Forum for Sustainable Development, Graz, Austria, 27 November.

Harrison, L. E. (1993) 'Underdevelopment is a State of Mind' in Seligson, M. A. and Passé-Smith, J. T. (eds) *Development and Underdevelopment: The Political Economy of Inequality* (London/Boulder: Lynne Rienner).

Harrison, R. J. (1974) *Europe in Question: Theories of Regional Integration* (London: George Allen and Unwin Ltd.).

Helleiner, G. K. (1996) 'Conventional Foolishness and Overall Ignorance: Current Approaches to Global Transformation and Development' in Jameson, K. P. and Wilber, C. K. (eds) *The Political Economy of Development and Underdevelopment* (New York: McGraw-Hill).

Herbert, R. (2002) 'Implications of the African Union and NEPAD for SADC' in *SADC-EU Relations: Looking Back and Moving Ahead* (Copenhagen: South African Institute of International Affairs and the Royal Danish Ministry of Foreign Affairs).

Hill, C. R. (1983) 'Regional Cooperation in Southern Africa', *African Affairs*, Volume 82, Number 327, April.

Hirst, P. and Thompson, G. (1999) *Globalisation in Question: The International Economy and the Possibilities of Governance*, 2nd edn (Cambridge: Polity Press).

Hirst, P. and Thomson, G. (2002) 'The Future of Globalisation', *Cooperation and Conflict*, Volume 37,Number 3, September.

Hoekman, B., Ng, F. and Olarreaga, M. (2001) 'Eliminating Excessive Tariffs on Exports of Least Developed Countries', World Bank Discussion Paper Number 2604.

Hoekman, B. M. and Kostecki, M. M. (2001) *The Political Economy of the World Trading System: The WTO and beyond*, 2nd edn (Oxford: Oxford University Press).

Holdren, J. P. and Smith, K. R. (2000) 'Energy, the Environment and Health' in Goldemberg, J. (ed.) *World Energy Assessment: Energy and the Challenge of Sustainability* (New York: UNDP).

Hollis, M. and Smith, S. (1990) *Explaining and Understanding International Relations* (Oxford: Clarendon Press).

Hooghe, L. (2001) *The European Commission and the Integration of Europe: Images of Governance* (Cambridge: Cambridge University Press).

Hooghe, L. and Marks, G. (2001) *Multi-Level Governance and European Integration* (Lanham: Rowman and Littlefield).

Hoogvelt, A. (1997) *Globalisation and the Postcolonial World: The New Political Economy of Development* (Basingstoke: Macmillan).

House of Commons Hansard Debates (2005) for 8 June (pt 6).

Howarth, R. (2000) 'The CAP: History and Attempts at Reform', *Economic Affairs*, Volume 20, Issue 2, June.

Howell, K. E. (1998) 'Neo-Functionalism Reassessed: Suggesting a Synthesis of European Integration Theory', School of Finance and Law (Bournemouth University) Working Paper Series Number 14.

Hughes, T. (2002) 'Poverty Eradication in SADC: Structural Challenges, Looming Crises' in *SADC-EU Relations: Looking Back and Moving Ahead* (Copenhagen: South African Institute of International Affairs and the Royal Danish Ministry of Foreign Affairs).

ICTSD (2007) 'EU Proposes Duty-and-Quota Free Access for ACP Countries in EPA Talks', *Bridges Weekly Trade News Digest*, Volume 11, Number 13, April.

Insausti, M. (2001) 'How Can Multilateral Trade Deliver Sustainable Development Outcomes?' (Part 2), WWF European Policy Office, Brussels, May.

International Centre for Trade and Sustainable Development (ICTSD) (2001) 'EU Modifies Everything But Arms Proposal under Pressure from Agricultural Sector', *Bridges Weekly Trade News Digest*, Volume 5, Number 3, 30 January.

International Court of Justice (1999) Summary of the Judgement of the International Court of Justice in the Matter of Kasikili/Sedudu Island – Botswana/Namibia, December.

International Rivers Network (2003) 'Human Rights Dammed off at Three Gorges: An Investigation of Resettlement and Human Rights Problems in the Three Gorges Dam Project', Report, January.

International Trade Forum (1999) 'Healthy Groundnuts: Controlling Aflatoxins', Trade Forum, Issue 2, 1 April.

Ion, S., Nicholls, D., Matzie, R. and Matzner, D. (2003) 'Pebble Bed Modular Reactor The First Generation IV Reactor To be Constructed', World Nuclear Association Annual Symposium, London, 3–5 September.

Isaksen, J. (2002) 'Restructuring SADC: Progress and Problems', Chr Michelsen Institute (CMI) Report, R2002: 15.

Isaksen, J. (2004) 'SADC in 2003: Restructuring and Progress in Regional Integration', Chr Michelsen Institute (CMI) Report, R2004:3.

Isaksen, J. and Tjønneland, E. N. (2001) 'Assessing the Restructuring of SADC: Positions, Policies and Progress', Chr Michelsen Institute (CMI) Report, R2001:6.

Jachtenfuchs, M. (2001) 'The Governance Approach to European Integration', *Journal of Common Market Studies*, Volume 39, Number 2, June.

Jachtenfuchs, M. and Kohler-Koch, B. (2004) 'Governance and Institutional Development' in Wiener, A. and Diez, T. (eds) *European Integration Theory* (Oxford: Oxford University Press).

Jackson, T. (1996) *Material Concerns: Pollution, Profit and Quality of Life* (London/ New York: Routledge).

Jacobs, M. (1991) *The Green Economy: Environment, Sustainable Development and the Politics of the Future* (London/Boulder: Pluto Press).

Jayaseela, F. (1996) 'Lesotho: Lesotho Highlands Water Project' in Nyborg, M. (ed.) *Power Conflicts* (Norway: Foreningen for Internasjonale Vann-og Skogstudier Report).

Jenkins, C. (2001) 'Integration and Cooperation in Southern Africa', OECD Development Centre Technical Papers, Number 172, March.

Jensen, M. F. (2002) 'Reviewing the SPS Agreement: A Developing Country Perspective', Centre for Development Research Working Paper 02.3, February.

Johansson, T. B. and Goldberg, J. (2002) 'Overview and a Policy Agenda' in Johansson, T. B. and Goldberg, J. (eds) *Energy for Sustainable Development: A Policy Agenda* (New York: UNDP).

Julian, M. (2007) 'EPA Update', *Trade Negotiation Insights*, May–June.

Kahn, H. (1993) 'The Confucian Ethic and Economic Growth' in Seligson, M. A. and Passé-Smith, J. T. (eds) *Development and Underdevelopment: The Political Economy of Inequality* (Boulder/London: Lynne Rienner).

Karingi, S., Lang, R., Oulmane, N., Perez, R., Jallab, M. and Hammouda, H. (2005) 'Economic and Welfare Impacts of the EU-Africa Economic Partnership Agreements', United Nations Economic Commission for Africa Concept Paper Number 10, March.

Keohane, R. and Hoffman, S. (1990) 'Conclusions: Community Politics and Institutional Change' in Wallace, W. (ed.) *The Dynamics of European Integration* (London/New York: Pinter Publishers).

Keohane, R. and Hoffman, S. (1994) 'Institutional Change in Europe in the 1980s' in Nelsen, B. F. and Stubb A. C-G. (eds) *The European Union: Readings on the Theory and Practise of European Integration* (Boulder/London: Lynne Rienner).

Khan, M. H. (2000) 'Rural Poverty in Developing Countries: Issues and Policies', IMF Working Paper, WP/00/78, April.

Kibreab, G. (2001) 'Property Rights, Development Policy and Depletion of Resources: The Case of the Central Highlands of Sudan, 1940s–1980s', *Environment and History*, Volume 7, Number 1.

Kirk, R. and Stern, M. (2003) 'The New Southern African Customs Union Agreement', World Bank Africa Region Working Paper Series Number 57, June.

Kirkpatrick, C. and Lee, N. (1999) 'WTO New Round – Sustainability Impact Assessment Study', Phase Two Report, Institute for Development Policy and Management, University of Manchester, 18 November.

Kishor, N., Mani, M. and Constantino, L. (2001) 'Economic and Environmental Benefits of Eliminating Log Export Bans – The Case of Costa Rica', IMF Working Paper WP/01/153, 2001.

Klare, M. (2003) 'The New Geopolitics', *Monthly Review*, Volume 55, Number 3, July–August.

Klunne, W., Cox, R., Gys, L., Purcell, C. and Louineau, J. P. (2002) 'Learning a Lesson: Assessing PV Programmes in Rural South Africa', *Renewable Energy World*, March–April.

Krimphoff, J. (2007) 'Sustainability Impact Assessment of the EU-ACP Economic Partnership Agreements: Summary of Key Findings, Policy Recommendations and Lessons Learned', PricewaterhouseCoopers.

Kritzinger-van Niekerk, L. and Pinto Moreira, E. (2002) 'Regional Integration in Southern Africa: Overview of Recent Developments', World Bank Regional Poverty Paper, December.

Kunzlik, P. (1994) *Environmental Policy* (Essex: Longman).

Kumar, P. (2001) 'Framework for Fair Trade and Poverty Eradication', Consumer Unity and Trust Society (CUTS) Centre for International Trade, Economics and Environment Briefing Paper Number 4.

Kuznets, S. (1993) 'Economic Growth and Income Inequality' in Seligson, M. A. and Passé-Smith, J. T. (eds) *Development and Underdevelopment: The Political Economy of Inequality* (Boulder/London: Lynne Rienner).

Kysar, D. A. (2003) 'Some Realism about Environmental Skepticism: The Implications of Bjørn Lomborg's *The Skeptical Environmentalist* for Environmental Law and Policy', *Ecology Law Quarterly*, Volume 30, Issue 223.

La Vina, A., Fransen, L., Faeth P. and Kurauchi, Y. (2006) 'Reforming Agricultural Subsidies: "No Regrets" Policies for Livelihoods and the Environment', World Resources Institute White Paper.

LDC Sugar Group (2007) 'Deliveries of EBA Sugar to the EU', 3 August, <http://www.ldcsugar.org>

Le Roux, C. J. B. (1999) 'The Botswana-Namibian Boundary Dispute in the Caprivi: To What Extent Does Botswana's Arms Procurement Program Represent a Drift towards Military Confrontation in the Region?' *Scientia Militaria*, Volume 29.

Lecomte, H -B. S. (2001) 'Effectiveness of Developing Country Participation in ACP-EU Negotiations', Overseas Development Institute Working Paper, October.

Lee, M. (2002) 'The European Union-South Africa Free Trade Agreement: In Whose Interest?' *Journal of Contemporary African Studies*, Volume 20, Issue 1.

Leestemaker, J. H. (2000) 'The Domino Effect, a Downstream Perspective in Water Management in Southern Africa' in Solomon, H. and Turton, A. (eds) *Water Wars: Enduring Myth or Impending Reality?* African Dialogue Monograph Series 2 (Accord: Durban).

Leith, J. C., Sackey, J. and Burns, D. (1999) 'Botswana: A Case Study of Economic Policy Prudence and Growth', World Bank Working Paper Number 20004, August.

Leitner, K. and Lester, S. (2004) 'WTO Dispute Settlement 1995–2003: A Statistical Analysis', *Journal of International Economic Law*, Volume 7, Issue 1.

Lewellen, T. C. (1995) *Dependency and Development* (Westport Connecticut/ London: Bergin & Garvey).

Lewis, J. D. (2001) 'Reform and Opportunity: The Changing Role and Patterns of Trade in South Africa and SADC', World Bank Africa Region Working Paper Series Number 14, March.

Lewis, J. D. (2002) 'Promoting Growth and Employment in South Africa', World Bank Africa Region Working Paper Series Number 32, June.

Lewis, J. D., Robinson, S. and Thierfelder, K. (2002) 'Free Trade Agreements and the SADC Economies', World Bank Africa Region Working Paper Series Number 27, February.

Leys, C. (1996) *The Rise and Fall of Development Theory* (Bloomington: Indiana University Press).

Liefferink, J. P., Lowe, P. D. and Mol, A. P. J. (1993) 'The Environment and the European Community: The Analysis of Political Integration' in Liefferink, J. P., Lowe, P. D. and Mol, A. P. J. (eds) *European Integration and Environmental Policy* (London: Belhaven Press).

Lomborg, B. (2001) *The Skeptical Environmentalist: Measuring the Real State of the World* (Cambridge: Cambridge University Press).

Lowe, I. (2003) 'Bjørn Lomborg Is Neither Sceptical Nor an Environmentalist', *Opinion*, 8, October.

Lynton-Evans, J. (1999) 'Complementarity and Coherence of EC Interventions in the Areas of Poverty Reduction, Food Security and Rural Development. Case Study: Mozambique.' Réseau Europeén de Sécurité Alimentaire (Resal) Technical Note Number NO4, June.

Mabey, N. and Arden-Clarke, C. (1999) 'Is There Really a North-South Split on Trade and Environment?' WWF Discussion Paper, November.

Madeley, J. (2000) *Hungry for Trade: How the Poor Pay for Free Trade* (London/New York: Zed Books).

Magdahl, J. (2006) 'Privatisation of Water – Public Private Partnerships: Do They Deliver to the Poor?' Report for the Norwegian Forum for Environment and Development, April.

Mail and Guardian (17 May 2004).

Malamud, A. (2001) 'Spillover in European and South American Integration: An Assessment', Latin American Studies Association, Conference Proceedings, Washington DC, 6–8 September.

Maliszewska, M. (2004) 'EU Enlargement: Benefits of the Single Market Expansion for Current and New Member States', Centre for Social and Economic Research, Studies and Analysis Paper Number 273, Warsaw, January.

Malzbender, D. (2003) 'Agricultural Reform in the EU: Implications of the June 2003 Common Agricultural Policy (CAP) – Reform for Southern Africa', Trade Law Centre for Southern Africa Working Paper Number 13/2003, November.

Mandelson, P. (2005) 'Raising Living Standards in Africa: The G8 Trade Challenge', Joint German Marshall Fund and Partnership to Cut Hunger and Poverty in Africa Event, Edinburgh, Speech/05/418, 6 July.

Martens, J. W., de Lange, T. J., Cloin, J., Szewczuk, S., Morris, R. and Zak, J. (2001) 'Accelerating the Market Penetration of Renewable Energy Technologies in South Africa: Action Plan Summary', Contract Number 4.1041/D/99-033 (ECN-C – 01-052), March.

Matjila, J. (2002) 'Lessons from EU-South Africa Negotiations', Conference Report, The Cotonou Trade Negotiations: Building European Solidarity for Southern Africa, ACP House, Brussels, 10 October.

Matthews, A. and Gallezot, J. (2006) 'The Role of EBA in the Political Economy of CAP Reform', IIIS Discussion Paper Number 133, April.

Mbeki, T. (2000) 'Integrating Africa into the World Economy', Africa-EU Summit, Cairo, 4 April.

Mbeki, T. (2003) *ANC Today*, 31 October.

McClelland, D. C. (1993) 'The Achievement Motive in Economic Growth' in Seligson, M. A. and Passé-Smith, J. T. (eds) *Development and Underdevelopment: The Political Economy of Inequality* (Boulder/London: Lynne Rienner).

McCulloch, N. and Ota, M. (2002) 'Export Horticulture and Poverty in Kenya', IDS Working Paper Number 174.

McDaniel, C. N. and Gowdy, J. M. (2000) *Paradise for Sale: A Parable of Nature* (Berkeley: University of California Press).

McDonald, D. A. (2002) 'No Money, No Service: South Africa's Citizens Lose Out under Attempts to Recover Service Costs for Water and Power', *Alternatives Journal*, Volume 28, Issue 2, Spring.

McQueen, M. (2002) 'EU Preferential Market Access Conditions for Least Developed Countries', *Intereconomics*, Volume 37, March–April.

Meier, G. M. (1995) *Leading Issues in Economic Development*, 6th edn (Oxford/New York: Oxford University Press).

Messerlin, P. A. (2001) *Measuring the Costs of Protection in Europe: European Commercial Policy in the 2000s* (Washington DC: Institute for International Economics).

Mill, J. S. (1994) *Principles of Political Economy and Chapters on Socialism* in Riley, J. (ed.) based on the 7th edn (1871) (Oxford: Oxford University Press).

Mitrany, D. (2003) 'A Working Peace System' in Nelsen, B. F. and Stubb, A. C -G. (eds) *The European Union: Readings on the Theory and Practice of European Integration*, 3rd edn (Boulder/London: Lynne Rienner).

Mkapa, B. (2004) Speech, Southern African Development Community Ordinary Summit, Mauritius, 16 August.

Montes, C., Wolfe, T., Cramer, C. and Hallam, A. (2000) *Evaluation of EC Country Strategy: Mozambique 1996–2000* (Brussels: EuropeAid Cooperation Office).

Moravcsik, A. (2003) 'Negotiating the Single European Act: National Interests and Conventional Statecraft in the European Community' in Nelsen, B. F.

and Stubb, A. C -G. (eds) *The European Union: Readings on the Theory and Practice of European Integration*, 3rd edn (Boulder/London: Lynne Rienner).

Morisset, J. (2000) 'Foreign Direct Investment in Africa: Policies Also Matter', World Bank Policy Research Working Paper Number 2481, November.

Morrison, J. A. and Pearce, R. (2003) 'Food Security and Trade: An Overview' in *Trade Reforms and Food Security: Conceptualising the Linkages* (Rome: FAO).

Morrissey, O. (2003) 'Trade Liberalisation and Food Security in Developing Countries' in *Trade Reforms and Food Security: Conceptualising the Linkages* (Rome: FAO).

Muir, J. (1991) *Our National Parks* (San Francisco: Sierra Club Books).

Murphy, S. (2003) 'The Role of Transnational Corporations' in *Trade Reforms and Food Security: Conceptualising the Linkages* (Rome: FAO).

Mutimer, D. (1994) 'Theories of Political Integration' in Michelmann, H. J. and Soldatos, P. (eds) *European Integration: Theories and Approaches* (Lanham/London: University Press of America).

Mutume, G. (2001) 'Africa Opposes New Round of Trade Talks', *African Recovery*, Volume 15, Number 3, October.

Muyakwa, S. L. (2001) 'Enhancing LDCs' Exports to OECD Markets: Challenges and Opportunities', African Resource Centre, Policy Brief, Number 2.

Naess, A. (1973) 'The Shallow and the Deep, Long-Range Ecology Movement: A Summary', *Inquiry*, Volume 1, Issue16.

Narlikar, A. (2005) 'Can Economic Integration Facilitate Regional Conflict Resolution?' CUTS International Briefing Paper, RECSA 4/2005.

Neumayer, E. (2001) 'Review: The Skeptical Environmentalist', *Times Higher Education Supplement*, 16 November.

Newark, J. W. (1995) 'Foreign Aid in the 1990s: The New Realities' in Dorraj, M. (ed.) *The Changing Political Economy of the Third World* (Boulder/ London: Lynne Rienner).

New Scientist (2006) 'A Future with No Bananas', 13 May.

Nicoll, W. and Salmon, T. C. (1994) *Understanding the New European Community* (New York/London: Harvester Wheatsheaf).

Nordström, H. and Vaughan, S. (1999) *World Trade Organisation Special Studies 4: Trade and Environment* (Geneva: WTO Publications).

Nuclear Science and Technology Research and Development (2001) Round-Table Discussion, January http://www.dst.gov.za/nuclearST.pdf

Obie, J. (2007) 'The Development of EBA' in Faber, G. and Obie, J. (eds) *European Union Trade Politics and Development: Everything But Arms Unravelled* (London/New York: Routledge).

O'Brien, R. and Williams, M. (2007) *Global Political Economy: Evolution and Dynamics*, 2nd edn (Basingstoke: Palgrave).

OCEANA (2006) 'United States Reasserts Commitment to Eliminating Subsidies Driving the Collapse of the World's Fisheries', Press Release, 28 February.

Organisation for Economic Cooperation and Development (OECD) (1998) *Globalisation and the Environment: Perspectives from OECD and Dynamic Non-Member Economies* (Paris: OECD Publications).

Ostergaard, T. (1990) *SADCC: A Political and Economic Survey* (Copenhagen: DANDIDA).

Otsuki, T., Wilson J. S. and Sewadeh, M. (2001) 'What Price Precaution: European Harmonisation of Aflatoxin Regulations and African Groundnut Exports', *European Review of Agricultural Economics*, Volume 28, Issue 2).

Oxfam (2003) 'EU Hypocrisy Unmasked: Why EU Trade Policy Hurts Development', Briefing Note, 5 May.

Page, S. (2002) 'The Nature of Negotiations with the EU: What Can We Learn from Other Trade Negotiations?' COMESA Seminar on EPAs, 31 October–1 November.

Page, S. and Bilal, S. (2001) 'Regional Integration in Western Africa', Overseas Development Institute Report, September.

Panagariya, A. (2005) 'Liberalising Agriculture', *Foreign Affairs*, WTO-Special Edn., Volume 84, Number 7, December.

Panagariya, A. (2006) 'Agricultural Liberalisation and the Least Developed Countries: Six Fallacies' in Greenaway, D. (ed.), *World Economy: Global Trade Policy 2005* (Oxford: Blackwell).

Panayotou, T. (1993) *Green Markets: The Economics of Sustainable Development* (San Francisco: ICS Press).

Pape, J. (2001) 'Poised to Succeed or Set up to Fail?' Municipal Services Project, Occasional Paper Series Number 1.

Parnell, S. (2000) 'Environment and Poverty in Southern Africa: Regional Linkages', Background Paper Prepared for DFID SA and CA.

Parsan, E. (2006) 'Aid for Trade: A Caribbean Perspective', Caribbean Regional Negotiating Machinery, Study Paper, May.

Passé-Smith, J. T. (1993) 'The Persistence of the Gap: Taking Stock of Economic Growth in the Post World War II Era' in Seligson, M. A. and Passé-Smith, J. T. (eds) *Development and Underdevelopment: The Political Economy of Inequality* (Boulder/London: Lynne Rienner).

Pauw, J. (2004) 'Metered to Death: How a Water Experiment Caused Riots and a Cholera Epidemic', Centre for Public Integrity, 5 April.

Pearce, D. (1992) 'Economics and the Global Environmental Challenge' in Markandya, A. and Richardson, J. (eds) *Environmental Economics* (London: Earthscan).

Pearce, D. (1993) *Economic Values and the Natural World* (London: Earthscan).

Pearce, D. W. and Turner, R. K. (1990) *Economics of Natural Resources and the Environment* (New York/London: Harvester Wheatsheaf).

Pearce, D., Turner, R. K. and Bateman, I. (1994) *Environmental Economics: An Elementary Introduction* (New York/London: Harvester Wheatsheaf).

Pearce, F. (2002) 'Giant Congo Hydroelectric Scheme a "Betrayal"', *New Scientist*, 30 September.

Pottinger, L. (2004) 'Can the Nile States Dam Their Way to Cooperation?', IRN Background Paper on the Nile Basin Initiative, March.

Practical Action (2005) 'Europe's Chance to Help Light up Africa: Energising Poverty Reduction', Practical Action Report, Rugby.

Prasad, G. and Visagie, E. (2005) 'Renewable Energy Technologies for Poverty Alleviation – Initial assessment report: South Africa', Energy Research Centre, University of Cape Town, South Africa, June.

Raffer, K. (2001) 'Cotonou: Slowly Undoing Lomé's Concept of Partnership', European Development Policy Study Group, Discussion Paper Number 21, Manchester Metropolitan University, October.

Rakner, L., van de Valle, N. and Mulaisho, D. (2002) 'Zambia' in *Aid and Reform in Africa: Lessons from Ten Case Studies* (Washington DC: World Bank Publication).

Randall, V. and Theobald, R. (1998) *Political Change and Underdevelopment: A Critical Introduction to Third World Politics*, 2nd edn (Basingstoke: Macmillan).

Raworth, K. (2002) 'The Great EU Sugar Scam: How Europe's Sugar Regime is Devastating Livelihoods in the Developing World', Oxfam Briefing Paper Number 27.

Reardon, T. (2003) 'Capital Market Liberalisation and the Latin American Agrifood System' in *Trade Reforms and Food Security: Conceptualising the Linkages* (Rome: FAO).

Reddy, A. K. N. (2002) 'Energy Technologies and Policies for Rural Development' in Johansson, T. B. and Goldemberg, J. (eds) *Energy for Sustainable Development: A Policy Agenda* (New York: UNDP).

Renewable Energy World (2001) 'Wind Energy in South Africa: Time to Implement', May–June.

Réseau Europeén de Sécurité Alimentaire (Resal) (1999) 'The "Everything but Arms" Initiative: What are its Consequences for the Resal Countries?' Technical Note.

Riddell, J. B. (1996) 'Things Fall Apart Again: Structural Adjustment Programmes in Sub-Saharan Africa' in Jameson K. P. and Wilber, C. K. (eds) *The Political Economy of Development and Underdevelopment* (New York: McGraw-Hill).

Ridley, M. (1995) *Down to Earth: A Contrarian View of Environmental Problems* (London: Institute of Economic Affairs).

Ridley, M. (1996) *Down to Earth II: Combating Environmental Myths* (London: Institute of Economic Affairs).

Rodney, W. (1982) *How Europe Underdeveloped Africa* (Washington DC: Howard University Press).

Rosamund, B. (2000) *Theories of European Integration* (Basingstoke: Macmillan).

Rosen, S. and Jones, W. (1979) 'The Radical Theory of Development' in Mack, A., Plant, A. and Doyle, U. (eds) *Imperialism, Intervention and Development* (London: Croon Helm).

Rostow, W. W. (1958) 'The Take-Off into Self-Sustained Growth' in Agarwala, A. N. and Singh, S. P. (eds) *The Economics of Underdevelopment* (Oxford/New York: Oxford University Press).

Sagoff, M. (1998) 'Carrying Capacity and Ecological Economics' in Crocker, D. A. and Linden, T. (eds) *Ethics of Consumption: The Good Life, Justice, and Global Citizenship* (Lanham, MD: Rowman and Littlefield).

Sandholtz, W. (1994) 'Choosing Union: Monetary Politics and Maastricht' in Nelsen, B. F. and Stubb, A. C-G. (eds) *The European Union: Readings on the Theory and Practise of European Integration* (Boulder/London: Lynne Rienner).

Sandholtz, W. and Zysman, J. (1994) '1992: Recasting the European Bargain' in Nelsen, B. F. and Stubb, A. C-G. (eds) *The European Union: Readings on the Theory and Practise of European Integration* (Boulder/London: Lynne Rienner).

Santoro, D. (2004) 'The "Aguas" Tango: Cashing in on Buenos Aires' Privatization', The Centre for Public Integrity.

Sarraf, M. and Jiwanji, M. (2001) 'Beating the Resource Curse: The Case of Botswana', World Bank Environment Department Papers, 24753, Number 1, October.

Sauvy, A. (1952) 'Three Worlds, A Planet', *L'Observateur*, Number 118, 15 August.

Sauvy, A. (1975) *Zero Growth?* (Oxford: Basil Blackwell).

Sawdon, G. (2002) 'Understanding and Responding to Fragile Livelihoods in Southern Africa', Save the Children, Report to the International Development Committee, October.

Schmidhuber, J. and Raney, T. (2003) 'Agricultural Trade, Trade Policies and the Global Food System' in Bruinsma, J. (ed.) *World Agriculture: Towards 2015/2030 – An FAO Perspective* (London: Earthscan).

Schmitter, P. C. (2004) 'Neo-Neofunctionalism' in Wiener, A. and Diez, T. (eds) *European Integration Theory* (Oxford: Oxford University Press).

Scholte, J. A. (2000) *Globalisation: A Critical Introduction* (Basingstoke: Macmillan).

Seligson, M. A. (1998) 'The Dual Gaps: An Updated Overview of Theory and Research' in Seligson, M. A. and Passé-Smith, J. T. (eds) *Development and Underdevelopment: The Political Economy of Global Inequality*, 2nd edn (Boulder/London: Lynne Rienner).

Sessions, G. and Devall, B. (1985) *Deep Ecology: Living as if Nature Mattered* (Utah: Perguine Smith).

Shafik, N. and Bandyopadhyay, S. (1992) 'Economic Growth and Environmental Quality: Time Series and Cross Country Evidence', World Bank Paper in Preparation for the World Development Report 1992.

Sharma, K. L. (2006) 'Food Security in the South Pacific Island Countries with Special Reference to the Fiji Islands', United Nations University Research Paper Number 2006/68, June.

Shiva, V. (1991) *The Violence of the Green Revolution: Third World Agriculture, Ecology and Politics* (London: Zed Books).

Shoprite Holdings LTD. (2007) < http://www.shoprite.co.za>

Simon, D. and Johnson, A. (1999) 'The Southern African Development Community: Regional Integration in Ferment', Royal Institute of International Affairs Briefing Paper, New Series Number 8, December.

Simon, J. (1981) *The Ultimate Resource* (Oxford: Martin Robertson).

Simon, J. (1996) *The Ultimate Resource II* (Princeton: Princeton University Press).

Simon, J. and Kahn, H. (1984) 'Introduction' in Simon, J. and Kahn, H. (eds) *The Resourceful Earth: A Response to Global 2000* (Oxford: Basil Blackwell).

Simon, J. and Wildavsky, A. (1984) 'On Species Loss, the Absence of Data, and Risks to Humanity' in Simon, J. and Kahn, H. (eds) *The Resourceful Earth: A Response to Global 2000* (Oxford: Basil Blackwell).

Smets, H. (2003) 'The Cost of Meeting the Johannesburg Targets for Drinking Water', French Water Academy Paper, October.

Smith, K. E. (2004) 'The ACP in the European Union's Network of Regional Relationships: Still Unique or Just One in the Crowd?' in Arts, K. and Dickson, A. (eds) *EU Development Cooperation* (Manchester: Manchester University Press).

South African Department of Trade and Industry (2004), 15 October, <http://www.thedti.gov.za/ econdb/resbank/rb6006SK.html>

South African Government (2004) 'Economy', 15 October, <http://www.info.gov.za/aboutsa/economy.htm#econ>

Southern African Development Community (SADC) (1995) *Protocol on Shared Watercourse Systems in the Southern African Development Community Region*, 28 August.

Southern African Development Community (SADC) (1998) *Regional Strategic Action Plan for Integrated Water Resources Development and Management in SADC Countries (1999–2004)*, SADC Water Sector Coordination Unit, Summary Report, September.

Southern African Development Community (SADC) (2003) *Food Security Crisis Update and Agro-Climatic Forecast as on 15th January 2003*, SADC Food Security Crisis Update, SADC Secretariat, 16 January.

Spalding-Fecher, R. (2002) *Sustainable Energy Watch: South Africa* (Paris: Helio International).

Stallings, B. (1992) 'International Influence on Economic Policy: Debt, Stabilisation and Structural Reform' in Haggard, S. and Kaufman, R. (eds) *The Politics of Economic Adjustment* (Princeton: Princeton University Press).

Steele, P., Le-Grand, S., Dobie, P., Hazelwood, P. and Bojö, J. (2002) *Linking Poverty Reduction and Environmental Management: Policy Challenges and Opportunities* (Washington DC: IBRD/World Bank).

Stevens, A. (1992) 'The Institutions of the European Community: To 1992 and beyond' in Dyker, D. (ed.) *The European Economy* (New York/London: Longman).

Stevens, C. (2003) 'Food Trade and Food Policy in Sub-Saharan Africa: Old Myths and New Challenges', *Development Policy Review*, Volume 21, Issue 5.

Stevens, C. (2006) 'The EU, Africa and Economic Partnership Agreements: Unintended Consequences of Policy Leverage', *Journal of Modern African Studies*, Volume 44, Issue, 3.

Stevens, C. (2007) 'Economic Partnership Agreements: What Happens in 2008', ODI Briefing Paper Number 23, June.

Stevens, C. and Kennan, J. (2001) 'The Impact of the EU's 'Everything but Arms' Proposal: A Report to Oxfam', Institute for Development Studies, January.

Stiglitz, J. (1998) 'More Instruments and Broader Goals: Moving towards the Post-Washington Consensus', World Institute for Development Economics Research Annual Lecture, Helsinki, 7 January.

Stop EPAs Coalition (2004) 'Six Reasons to Oppose EPAs in their Current Form', Position Paper, November.

Strauss, A. and Corbin, J. (1990) *Basics of Qualitative Research: Grounded Theory Procedures and Techniques* (California: Sage).

Stuart, L. (2005) 'Truth or Consequences: Why the EU and the USA Must Reform their Subsidies, or Pay the Price', Oxfam Briefing Paper 81, November.

Sturm, D. (1999) 'Product Standards and "Green Protectionism"', Centre for Economic Performance, London School of Economics and Political Science, 10 April.

Suez Water (2007) 'Public-Partnership in Management of Water Services: AnEffective Approach', <http://www.suez-environnement.com/en/engagements/ access-to-water/water-for-all/water-for-all>

Sutcliffe, A. (1996) *An Economic and Social History of Western Europe Since 1945* (London/New York: Longman).

Thatcher, M. (1993) *The Downing Street Years, 1979–1990* (New York: HarperPerennial).

Theodorakis, A. (2002) 'The EU Energy Initiative', Speech to DG Development Stakeholder Meeting, 23 July.

Tierney, J. (1990) 'Betting the Planet', *The New York Times Magazine*, 2 December.

Trade Negotiation Insights (2004) Volume 3, Number 1, January.

Tucker, R. P. (2000) *Insatiable Appetite: The United States and the Ecological Degradation of the Tropical World* (Berkeley: University of California Press).

Turner, M. and Hulme, D. (1997) *Governance, Administration and Development: Making the State Work* (Connecticut: Kumarian Press).

Turton, A. R. (2000) 'Water Wars in Southern Africa: Challenging Conventional Wisdom' in Solomon, H. and Turton, A. (eds) *Water Wars: Enduring Myth or Impending Reality?* African Dialogue Monograph Series 2 (Durban: Accord).

United Nations (2005) Millennium Development Goals, <http://www.un.org/millenniumgoals>

United Nations Development Programme (UNDP) (2006) *Human Development Report 2006 – Beyond Scarcity: Power, Poverty and the Global Water Crisis* (Basingstoke: Palgrave Macmillan).

United Nations Development Programme (UNDP) (2007) 'Globalisation and the Least Developed Countries', Conference Background Paper, Making Globalisation Work for the LDCs, Istanbul, 9–11 July.

United Nations Economic Commission for Africa (UNECA) (2005a) 'Assessment of the Impact of the Economic Partnership Agreement between the COMESA Countries and the European Union', September.

United Nations Economic Commission for Africa (UNECA) (2005b) 'Assessment of the Impact of the Economic Partnership Agreement between the ECOWAS Countries and the European Union', Draft Paper , December.

United Nations Office for the Coordination of Humanitarian Affairs (2003) 'Tanzania: Focus on Impact of Water Reforms on Dar es Salaam's Poor', Press Release, 5 December.

United Nations Office of the High Representative for the Least Developed Countries, Landlocked Developing Countries and Small Island Developing Countries, 'The Criteria for the Identification of the LDCs', <http://www.un.org/special-rep/ohrlls/ldc/ldc%20criteria.htm>

United Nations Office of the High Representative for the Least Developed Countries, Landlocked Developing Countries and Small Island Developing Countries, 'List of Least Developed Countries', <http://www.un.org/special-rep/ohrlls/ldc/list.htm>

UNESCO (2004) Press Release, 20 January, <http://www.wateryear2003.org/en/ev.php-URL_ID=6427&URL_DO=DO_TOPIC&URL_SECTION=201.html>

United States Department of Agriculture (USDA) (2006) Economic Research Service, 'Real Per Capita Income (2000 base) – Historical', updated 19 December, <http://www.ers.usda.gov/data/macroeconomics/#HistoricalMacro Tables>

Unnevehr, L. J. (2000) 'Food Safety Issues and Fresh Product Exports from LDCs', *Agricultural Economics*, Volume 23, Issue 3, September.

Valenzuela, J. S. and Valenzuela, A. (1993) 'Modernisation and Dependency: Alternative Perspectives in the Study of Latin American Underdevelopment' in Seligson, M. A. and Passé-Smith, J. T. (eds) *Development and Underdevelopment: The Political Economy of Inequality* (Boulder/London: Lynne Rienner).

Van Gelder, J. W., Van der Valk, F., Dros, J. M. and Worm, J. (2002) 'The Impacts and Financing of Large Dams', WWF International Research Paper (Living Waters Programme), November.

Van Wyk, J. (1998) 'Towards Water Security in Southern Africa', *African Security Review*, Volume 7, Number 2.

Vaz, A. C. and Van der Zaag, P. (2003) 'Sharing the Incomati Waters: Cooperation and Competition in the Balance', UNESCO Technical Documents in Hydrology, PC-CP Series Number 14, SC-2003/WS/46.

Wade, H. (2005) 'Demonstration Projects to Showcase the Business Angle of Renewable Energy Service Delivery in the Pacific Islands', Pacific Islands Renewable Energy Project Report, Pacific Regional Environment Programme.

Watkins, K. (2004) 'Dumping on the World: How EU Sugar Policies Hurt Poor Countries', Oxfam Briefing Paper Number 61, March.

Watkins, K. (2005) *Human Development Report 2005 – International Cooperation at a Crossroads: Aid, Trade and Security in an Unequal World* (New York: UNDP).

Watkins, K. and Fowler, P. (2004) 'Rigged Rules and Double Standards: Trade, Globalisation, and the Fight Against Poverty', Oxfam Campaign Report, February.

Welsch, H. (1996) 'Recycling of Carbon/Energy Taxes on the Labour Market', *Environmental and Resource Economics*, Volume 8.

Williamson, J. (2004) 'The Washington Consensus as Policy Prescription for Development', World Bank Practitioners of Development Lecture, 13 January.

Wilson, J. S. (2001) 'Advancing the WTO Agenda on Trade and Standards: A Developing Country Voice in the Debate', The African Economic Research Consortium Conference on Trade, Conference Paper, Geneva, 8–9 March.

World Bank (1981) *Accelerated Development in Sub-Saharan Africa: An Agenda for Action* (Washington DC: World Bank).

World Bank (1992) *World Development Report 1992 – Development and the Environment* (Oxford/New York: Oxford University Press).

World Bank (2001) 'Aid and Reform in Africa', Regional Report, 27 March.

World Bank (2004) 'World Development Indicators 2004', <http://iresearch.worldbank.org/PovcalNet/jsp/index.jsp>

World Bank (2005a) 'World Bank and Partners Launch Initiative to "Turn the Tide" of Fisheries Depletion', Press Release, 24 August.

World Bank (2005b) 'Mozambique: Data and Statistics', April, <http://web.worldbank.org/WBSITE/EXTERNAL/COUNTRIES/AFRICAEXT/MOZAMBIQUEEXTN/0,,menuPK:382158~pagePK:141132~piPK:141109~theSitePK:382131,00.html>

World Bank (2006a) *Africa Development Indicators 2006* (Washington DC: World Bank).

World Bank (2006b) 'Debt Relief for the Poorest: An Evaluation Update of the HIPC Initiative' (Washington DC: World Bank).

World Commission on Dams (2000) *Dams and Development: A New Framework for Decision Making* (Earthscan: London).

World Commission on Environment and Development (1987) *Our Common Future* (The 'Bruntland Commission') (Oxford University Press: Oxford).

World Health Organisation (2004) 'Water, Sanitation and Hygiene Links to Health: Facts and Figures: Facts and Figures', Water, Sanitation and Health, March.

World Hydrological Cycle Observing System (WHYCOS) (2007) Home Page, <http://www.wmo.ch/web/homs/projects/whycos.html>

World Trade Organisation (2004) *Report on the Activities of the Committee on Sanitary and Phytosanitary Measures*, G/L/709, 28 October.

World Trade Organisation (2007) 'Activities of the SPS Committee and other Relevant WTO Activities from 2006 to the Present', Report for the Food and Agriculture Organisation/World Health Organisation Food Standard Programme, Rome, 2–7 July.

Yari, M., ESCAP Secretariat and Duncan, R. (2004) 'Food Security: A Development Issue for Pacific Island Countries' in Kumar, R. (ed.) *UNESCAP Bulletin on Asia-Pacific Perspectives 2004–2005*, (New York: United Nations).

Yu, W. and Jensen, T. (2003) 'Trade Preferences, WTO Negotiations and the LDCS: The Case of the "Everything but Arms" Initiative', Danish Research Institute of Food Economics, 30 April.

Zafar, A. and Kubota, K. (2003) 'Regional Integration in Central Africa: Key Issues', World Bank Africa Region Working Paper Number 52, June.

Ziegler, J. (2006) Economic, Social and Cultural Rights – The Right to Food UNESC Report E/CN.4/2006/44/Add.2.

Zieroth, G. (2005) 'Pacific Islands Energy Policy and Strategic Action Planning – Annual Report 2005', South Pacific Applied Geoscience Commission, <http://www.sopac.org/data/virlib/PI/PI0041.pdf>

Zuidberg, J. J. (1995) 'Lesotho-EU Cooperation: Power for the People', *The ACP-EU Courier*, Number 154, November–December.

Index